TABLE OF CONTENTS

TO THE TEACHER

THE BIG SUBTRACTION BOOK IS A COLLECTION OF FUN-FILLED ACTIVITIES TO INTRODUCE AND REINFORCE BASIC SUBTRACTION FACTS. IT EMPHASIZES THE USE OF NUMBER PAIRS WITH THE SAME DIFFERENCE. FOR EXAMPLE: 7 - 3 = 4, 6 - 2 = 4, AND 5 - 1 = 4.

AT THE END OF THIS BOOK THERE IS A THREE PAGE TEST THAT CAN BE USED TO TEST AND RETEST EACH STUDENT ON THE BASIC FACTS. WHEN THE STUDENT HAS COMPLETED THE TEST, HE/SHE CAN GRAPH THE SCORE. THEN THE TEST CAN BE CUT APART AND USED AS A SET OF FLASH CARDS.

THE BOOK INCLUDES SEVERAL AWARDS TO FURTHER ACHIEVEMENT AND HARD WORK.

THE BOOK IS ARRANGED WITH THE EASIEST FACTS INTRODUCED FIRST. IF A STUDENT HAS A DIFFICULT TIME WITH THE BASIC FACTS IN THE FIRST SECTION, ADDITIONAL WORKSHEETS COVERING THE SAME FACTS SHOULD BE GIVEN BEFORE THE STUDENT IS ALLOWED TO CONTINUE ON TO THE NEXT SECTION.

CLASSROOM COMPETITION GAMES SUCH AS MATH BASEBALL OR MATH BINGO CAN ALSO BE USED TO HELP REINFORCE THE BASIC FACTS. MATH BASEBALL IS PLAYED BY DIVIDING THE CLASS INTO TWO TEAMS. THE TEACHER OR ONE STUDENT SHOULD BE THE PITCHER AND SHOULD SHOW THE PLAYER OF THE RECEIVING TEAM A FLASHCARD. THE BATTER (RECEIVER) SHOULD BE ALLOWED NO MORE THAN 5 SECONDS TO ANSWER CORRECTLY. IF CORRECT THE PLAYER MAY ADVANCE TO FIRST BASE. IF INCORRECT, THE TEAM HAS ONE OUT. THREE OUTS, OF COURSE, AND IT IS THE OPPOSING TEAMS TURN TO BAT. FOUR CORRECT ANSWERS WILL RESULT IN A RUN SCORED FOR THE BATTING TEAM.

TO PLAY MATH BINGO, MAKE BINGO TYPE CARDS WITH DIFFERENT ANSWERS IN THE VARIOUS SPACES. NUMBERS USED CAN RANGE FROM 0 TO 25, OR 25 TO 50. THE TEACHER FLASHES A CARD TO THE PLAYERS. IF A PLAYER HAS THAT ANSWER ON HIS/HER CARD, HE/SHE MAY COVER THAT SPOT. WINNERS ARE DETERMINED AS IN ANY BINGO STYLE GAME.

SUBTRACTION

2+1=3

3-1=2

3-2=1

WRITE 2 SUBTRACTION PROBLEMS FOR EACH OF THE FOLLOWING ADDITION PROBLEMS.

3+1=4

____

____

2+3=5

____

____

4+2=6

____

____

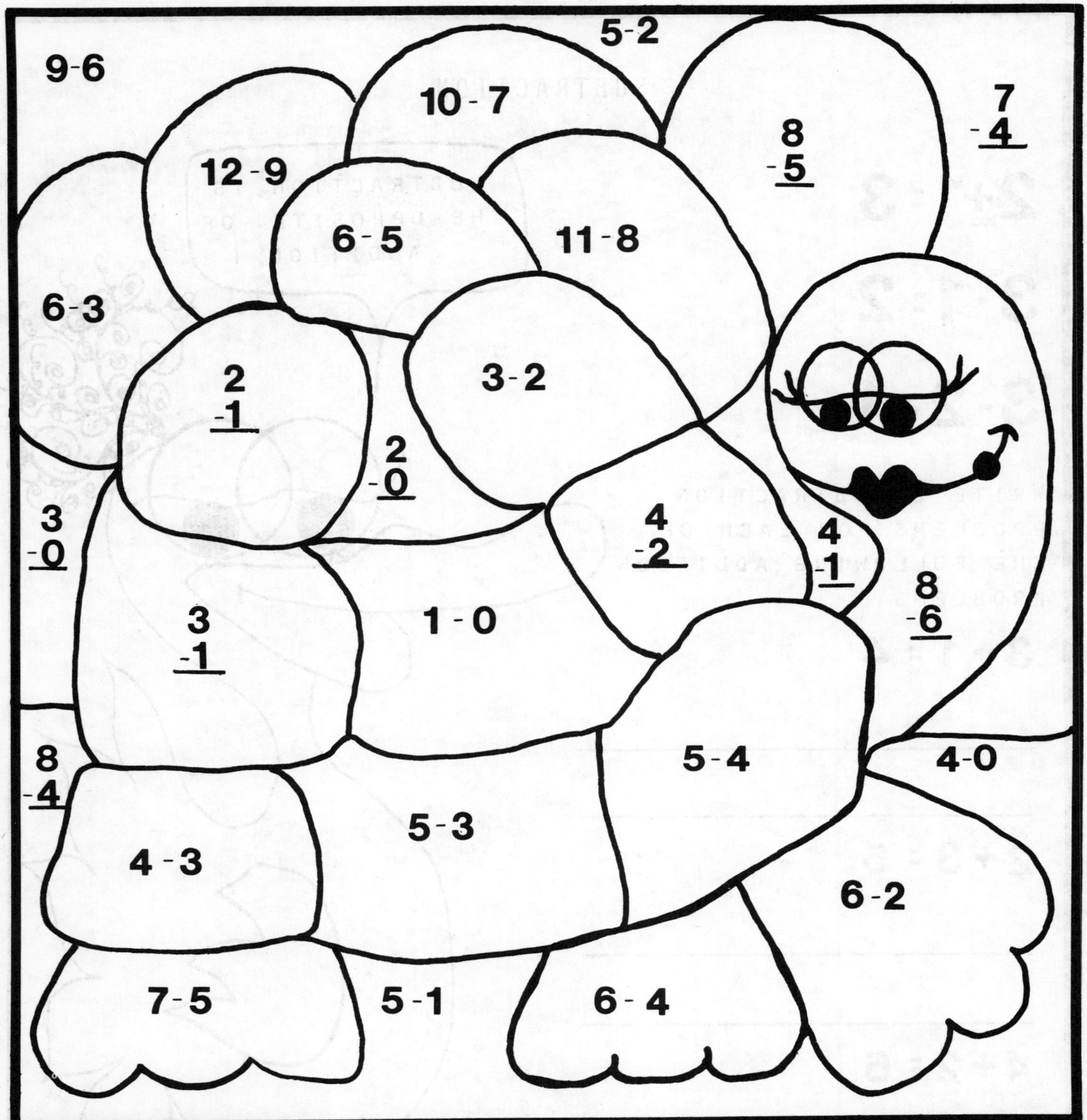

DIRECTIONS:

IF THE DIFFERENCE IS 1, COLOR THE AREA PINK.

IF THE DIFFERENCE IS 2, COLOR THE AREA PURPLE.

IF THE DIFFERENCE IS 3, COLOR THE AREA GREEN.

IF THE DIFFERENCE IS 4, COLOR THE AREA YELLOW.

CIRCLE ALL THE NUMBER PAIRS WITH A DIFFERENCE OF 1, RED. CIRCLE ALL THE NUMBER PAIRS WITH A DIFFERENCE OF 2, GREEN.

7	8	6	9	7	7	5
5	6	4	7	5	4	5
4	8	0	5	3	1	8
8	6	5	3	5	4	6
6	4	5	0	8	0	9
7	8	4	5	6	1	7
5	9	7	1	4	5	5

* THE LARGE NUMBER MUST COME FIRST

CIRCLE THE NUMBER PAIRS WITH A DIFFERENCE OF 3, RED.
CIRCLE THE NUMBER PAIRS WITH A DIFFERENCE OF 4, YELLOW.

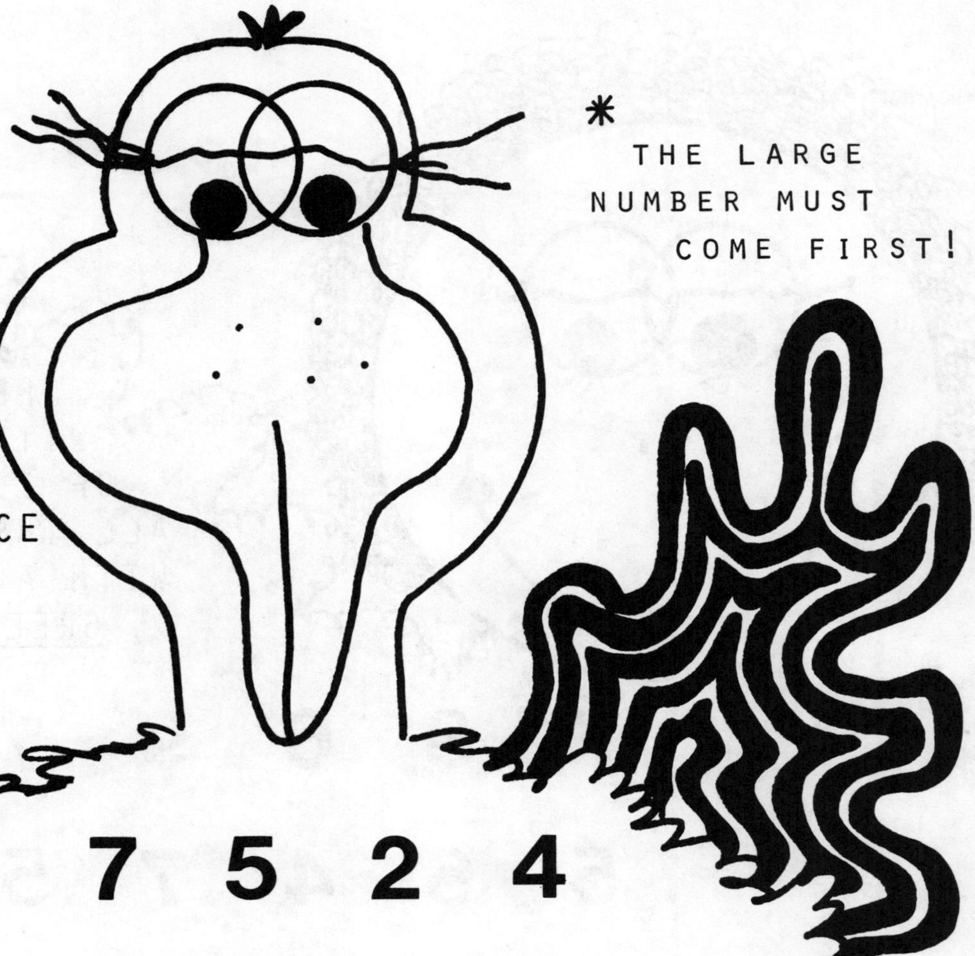

* THE LARGE NUMBER MUST COME FIRST!

7 5 2 4

8 4 3 6 2 1 3

5 9 5 5 9 4 0

6 7 4 1 6 9 7

3 0 4 8 4 6 4

9 6 1 5 8 7 8

5 3 0 1 5 4 4

DIRECTIONS: WITH SCISSORS CUT APART THE EQUATIONS AT THE BOTTOM OF THE PAGE. GLUE EACH EQUATION IN THE CORRECT BOX.

1	2
3	4

6 - 5	6 - 4	5 - 4
7 - 5	7 - 3	8 - 5
8 - 4	9 - 5	7 - 4
6 - 3	5 - 3	8 - 6
7 - 6	8 - 7	9 - 8
9 - 7	9 - 6	6 - 2

MAGIC SQUARES

DIRECTIONS: THE DIFFERENCE BETWEEN THE FIRST TWO NUMBERS IS THE LAST NUMBER IN EACH ROW AND COLUMN. CAN YOU FIND THE MISSING NUMBERS?

7		
		0
5		4

6		4
	0	
4		

4		0
1		
		0

6		
	0	4
2		

MORE MAGIC SQUARES

4	1	
2		
		2

6	3	
3		
	0	

7	2	
6		
		0

6	3	
		1
	0	

3	0	
		0
3		

9	5	
6		
		2

HIDDEN CIRCLES

COLOR THE AREAS WITH A DIFFERENCE OF 1, GREEN.

COLOR THE AREAS WITH A DIFFERENCE OF 2, PINK.

COLOR THE AREAS WITH A DIFFERENCE OF 3, ORANGE

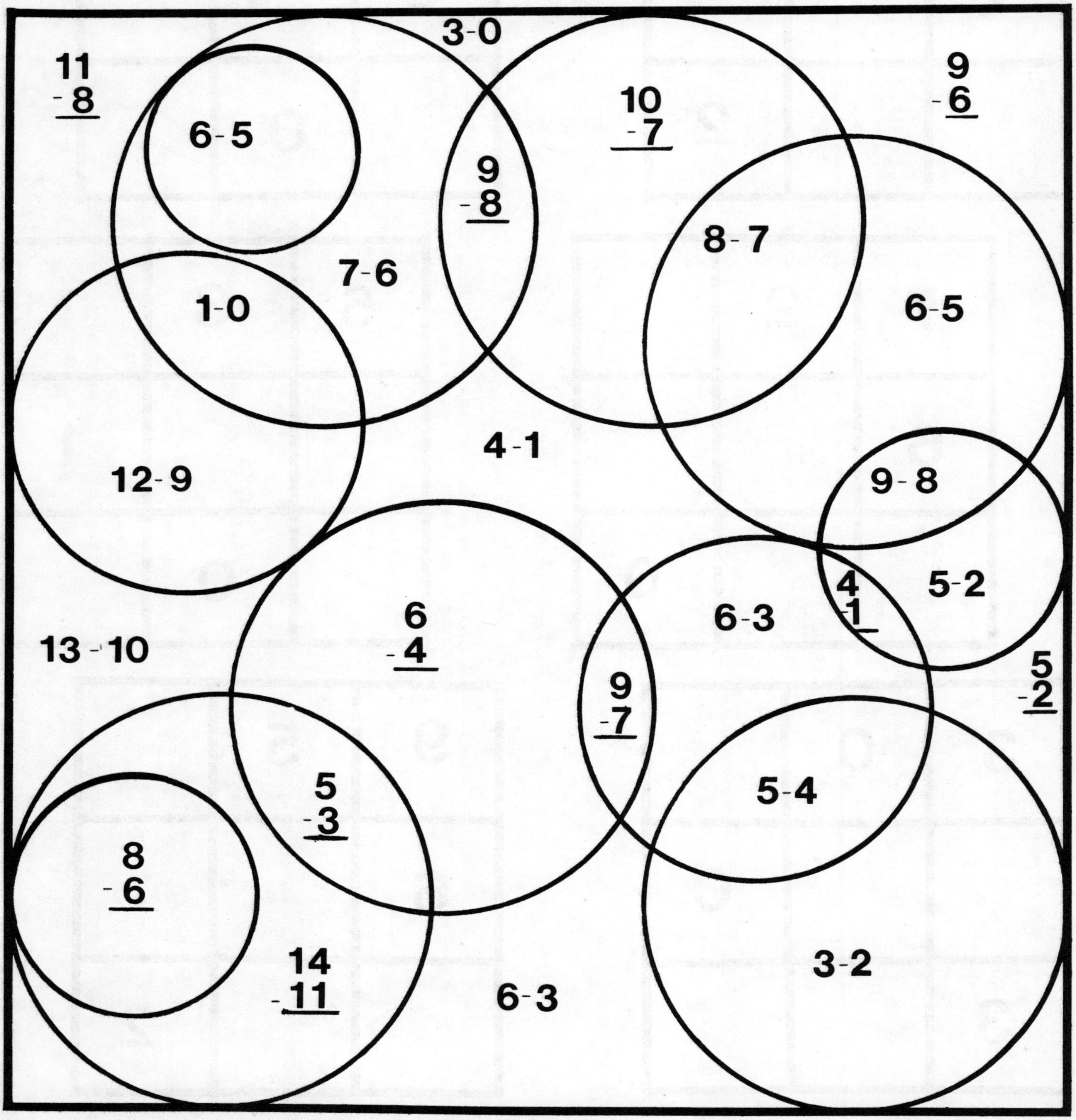

SATELLITE SUBTRACTION

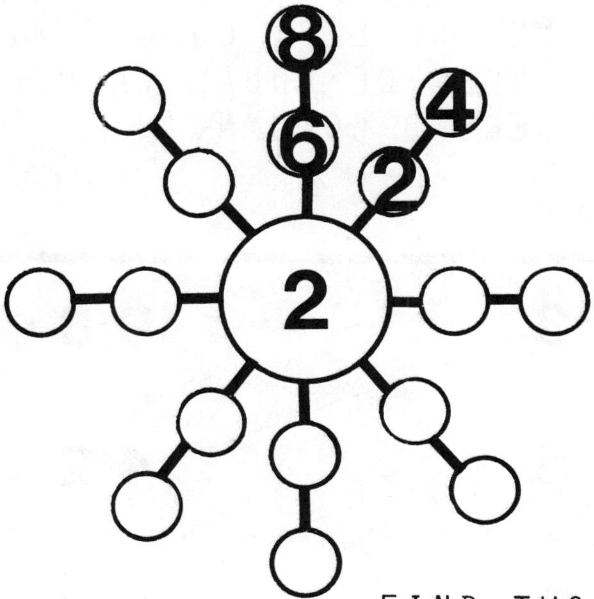

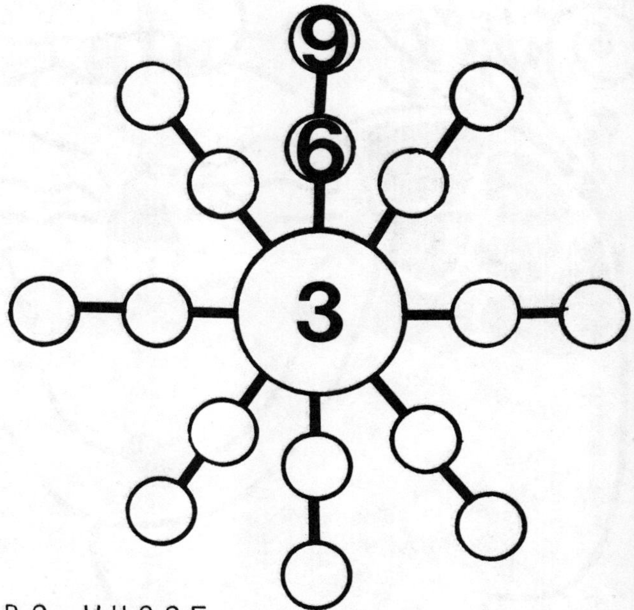

FIND TWO NUMBERS WHOSE DIFFERENCE EQUALS THE NUMBER IN THE CENTER OF EACH SATELLITE. WRITE IN ALL MISSING NUMBERS.

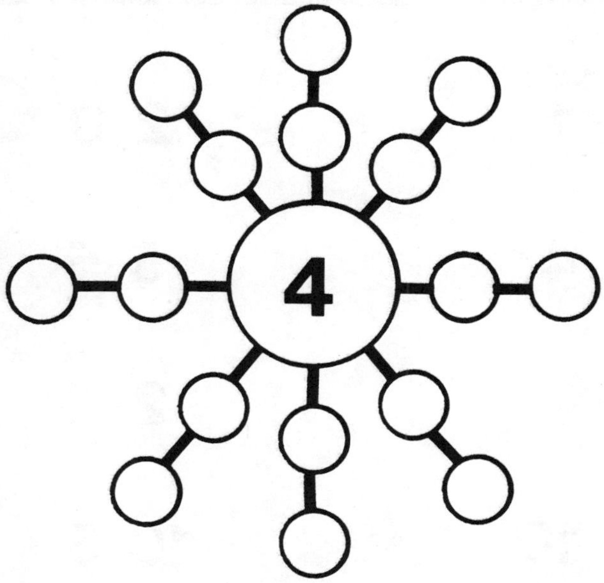

MATCHING DIFFERENCES

DIRECTIONS:

MATCH A SUBTRACTION PROBLEM IN THE LEFT COLUMN WITH A PROBLEM OF EQUAL VALUE IN THE SECOND COLUMN.

5-3 9-8

9-5 4-2

7-6 6-3

5-2 4-0

8-8 9-9

7-6 1-0

9-5 10-5

5-0 8-4

4-2 7-6

8-7 6-4

1-1 2-0

6-4 5-5

7-4 2-1

15-10 10-5

9-8 6-3

CIRCLE THE SUBTRACTION PROBLEMS THAT ARE FOLLOWED BY THEIR ANSWERS.

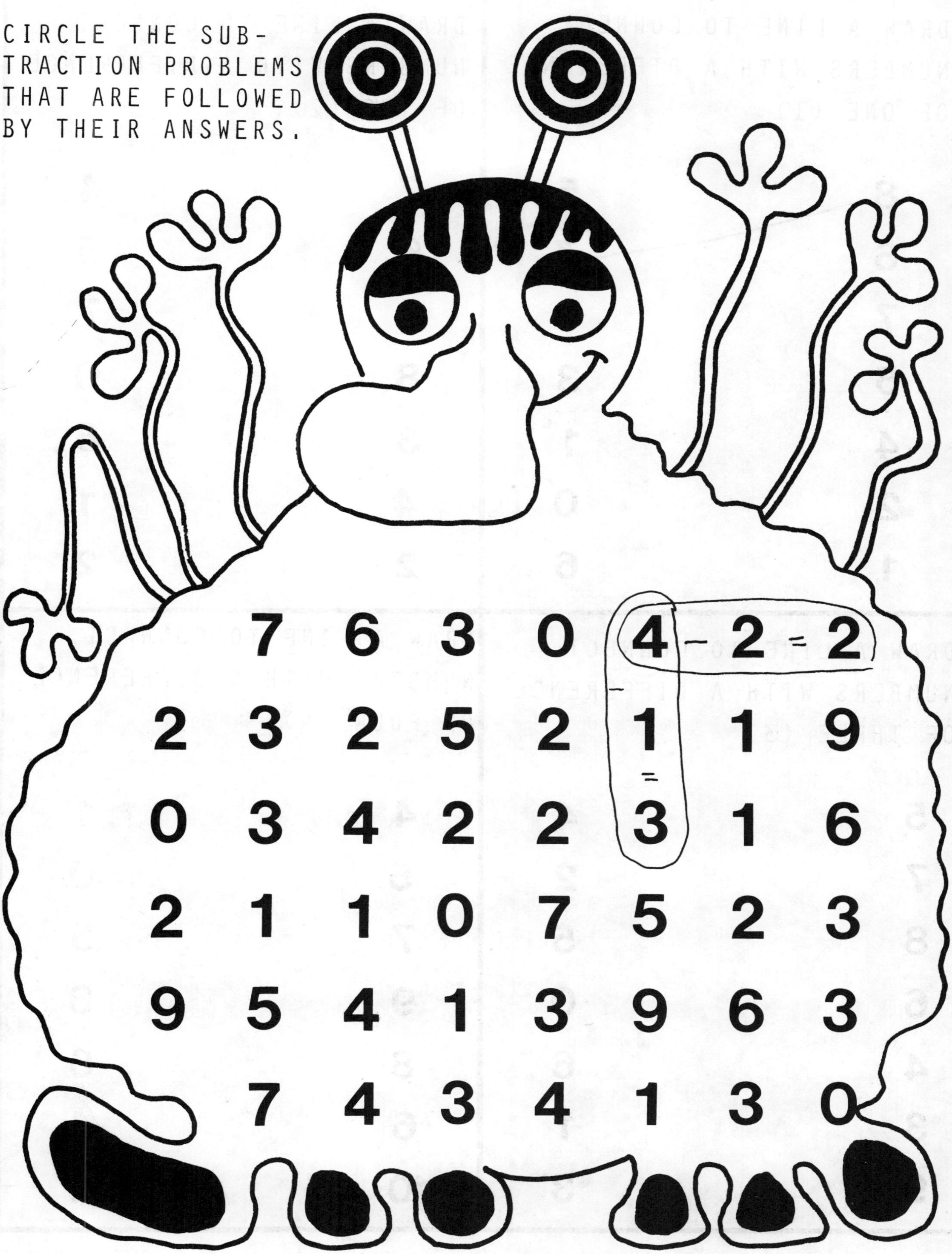

	7	6	3	0	4	2	2
2	3	2	5	2	1	1	9
0	3	4	2	2	3	1	6
2	1	1	0	7	5	2	3
9	5	4	1	3	9	6	3
	7	4	3	4	1	3	0

DRAW A LINE TO CONNECT NUMBERS WITH A DIFFERENCE OF ONE (1).

8	5
6	7
7	4
5	3
4	1
2	0
1	6

DRAW A LINE TO CONNECT NUMBERS WITH A DIFFERENCE OF TWO (2).

7	4
6	5
9	7
8	0
3	6
4	1
2	2

DRAW A LINE TO CONNECT NUMBERS WITH A DIFFERENCE OF THREE (3)

5	4
7	2
8	5
6	0
4	6
3	1
9	3

DRAW A LINE TO CONNECT NUMBERS WITH A DIFFERENCE OF FOUR (4).

4	1
5	0
7	5
9	3
8	6
6	4
10	2

START IN THE CIRCLES MARKED WITH THE NUMBER TEN (10). MOVE ALONG THE PATH CONNECTING THE CIRCLES. SUBTRACT EACH NUMBER AS YOU PASS THROUGH THE CIRCLE. WRITE THE DIFFERENCE IN THE EMPTY CIRCLE AT THE END.

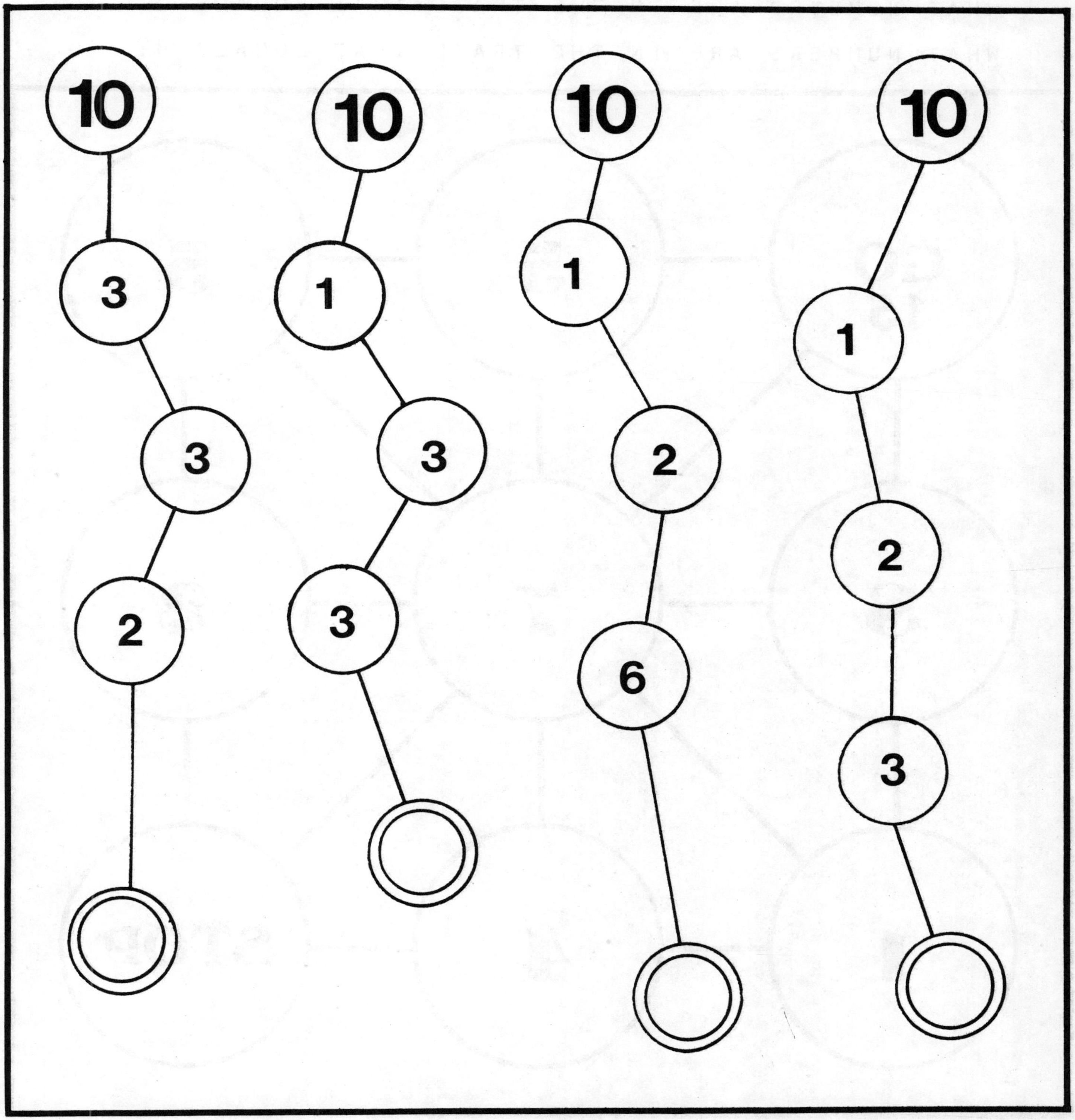

BEGIN AT THE "GO". YOU HAVE 15 POINTS. AS YOU TRAVEL THROUGH A CIRCLE, SUBTRACT "THAT" MANY POINTS.

WHAT NUMBERS ARE IN THE TRAIL THAT EQUALS 1?

WHAT NUMBERS ARE IN THE TRAIL THAT EQUALS 2?

WHAT NUMBERS ARE IN THE TRAIL THAT EQUALS 3?

WHAT NUMBERS ARE IN THE TRAIL THAT EQUALS 4?

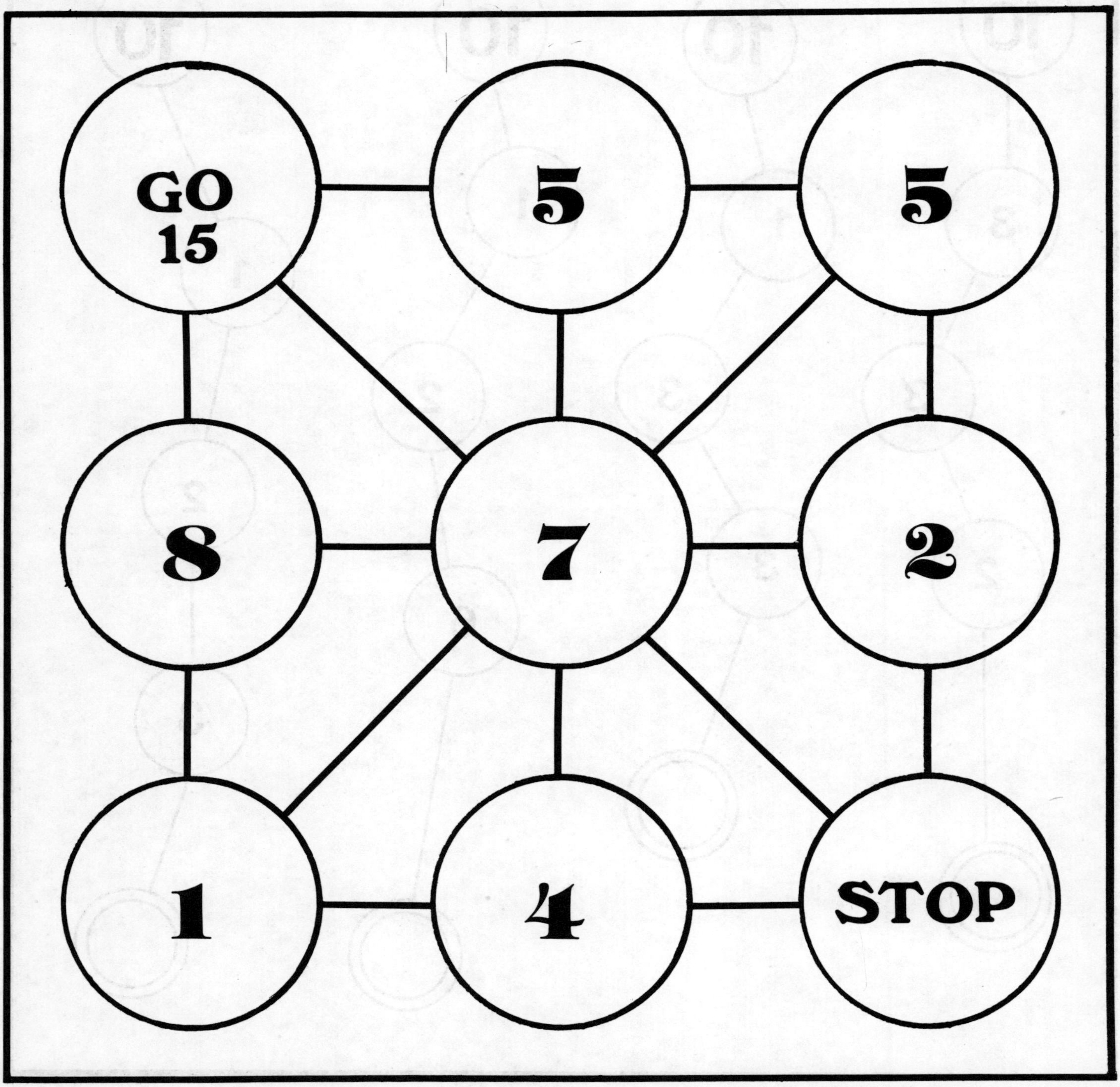

LINE PUZZLES

CAN YOU DRAW 2 STRAIGHT LINES THAT WILL DIVIDE THE SQUARE SO THAT EACH AREA HAS 2 NUMBERS WHOSE DIFFERENCE IS 3?

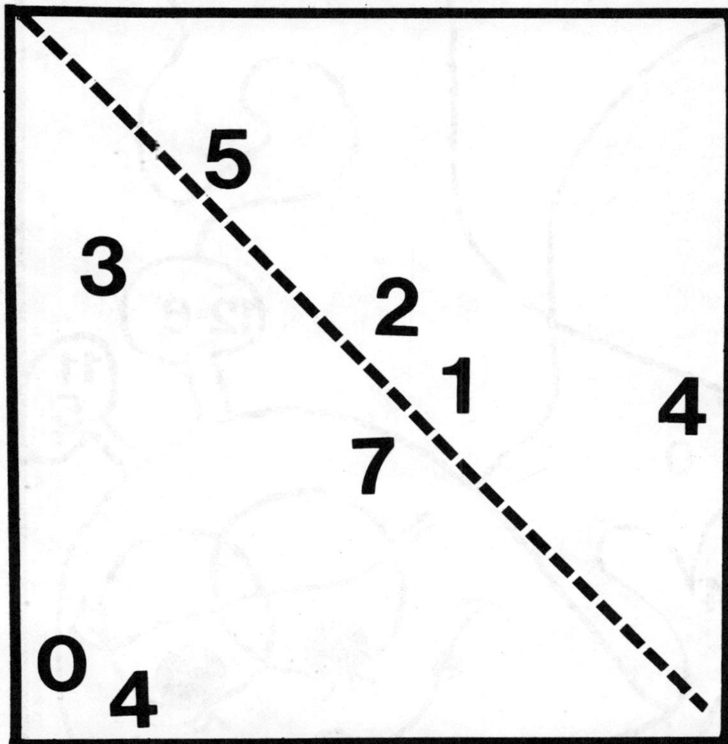

CAN YOU DRAW 2 STRAIGHT LINES THAT WILL DIVIDE THE SQUARE SO THAT EACH AREA HAS 2 NUMBERS WHOSE DIFFERENCE IS 4?

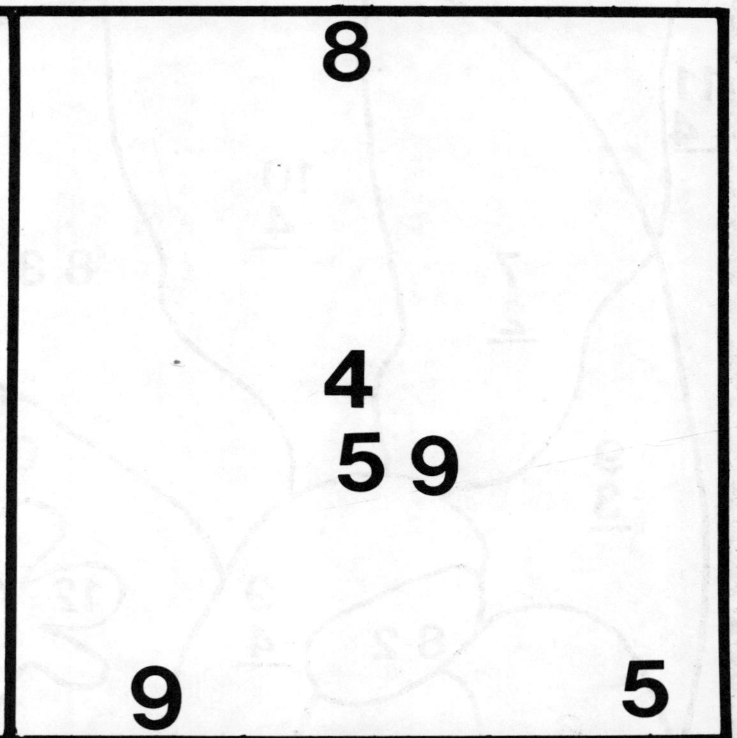

CAN YOU DRAW 2 STRAIGHT LINES THAT WILL DIVIDE THE SQUARE SO THAT EACH AREA HAS 2 NUMBERS WHOSE DIFFERENCE IS 1?

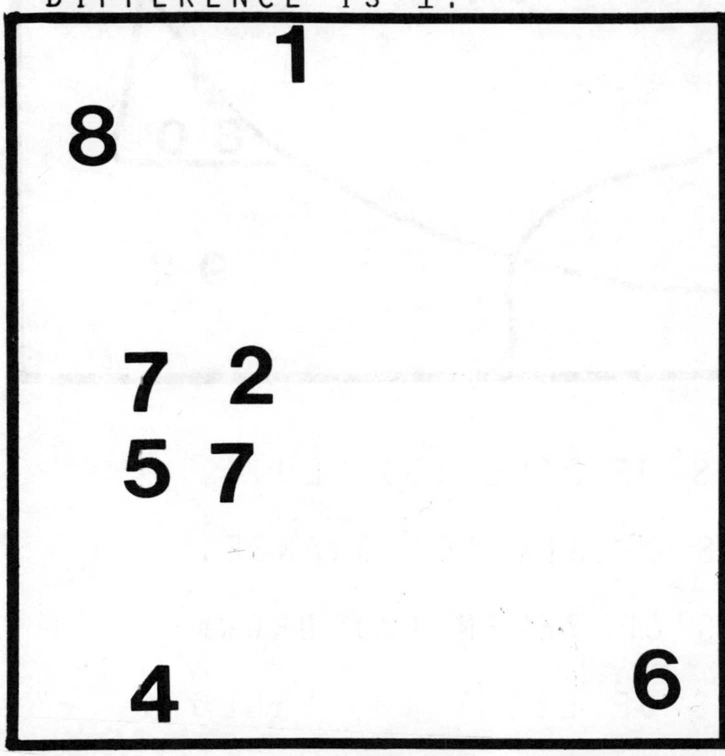

CAN YOU DRAW 2 STRAIGHT LINES THAT WILL DIVIDE THE SQUARE SO THAT EACH AREA HAS 2 NUMBERS WHOSE DIFFERENCE IS 2?

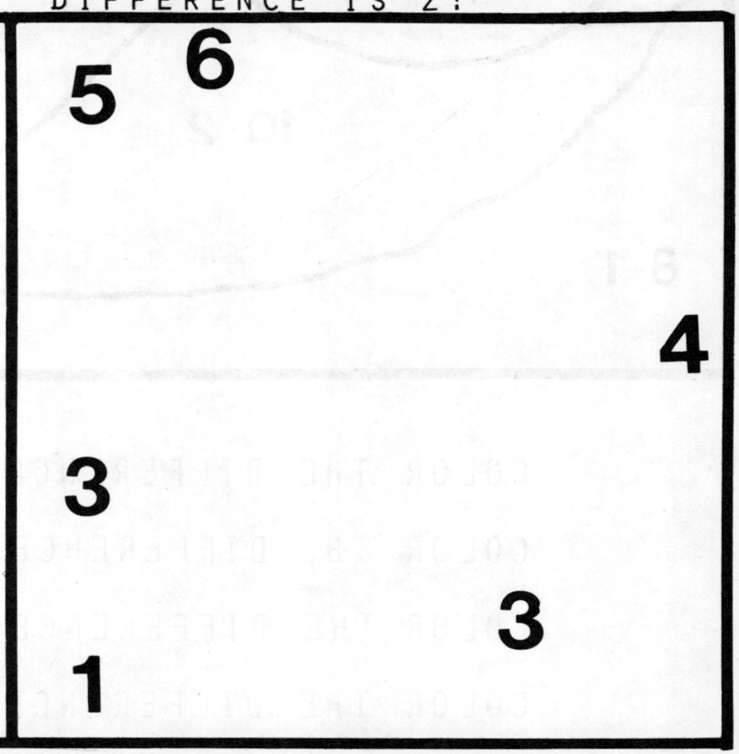

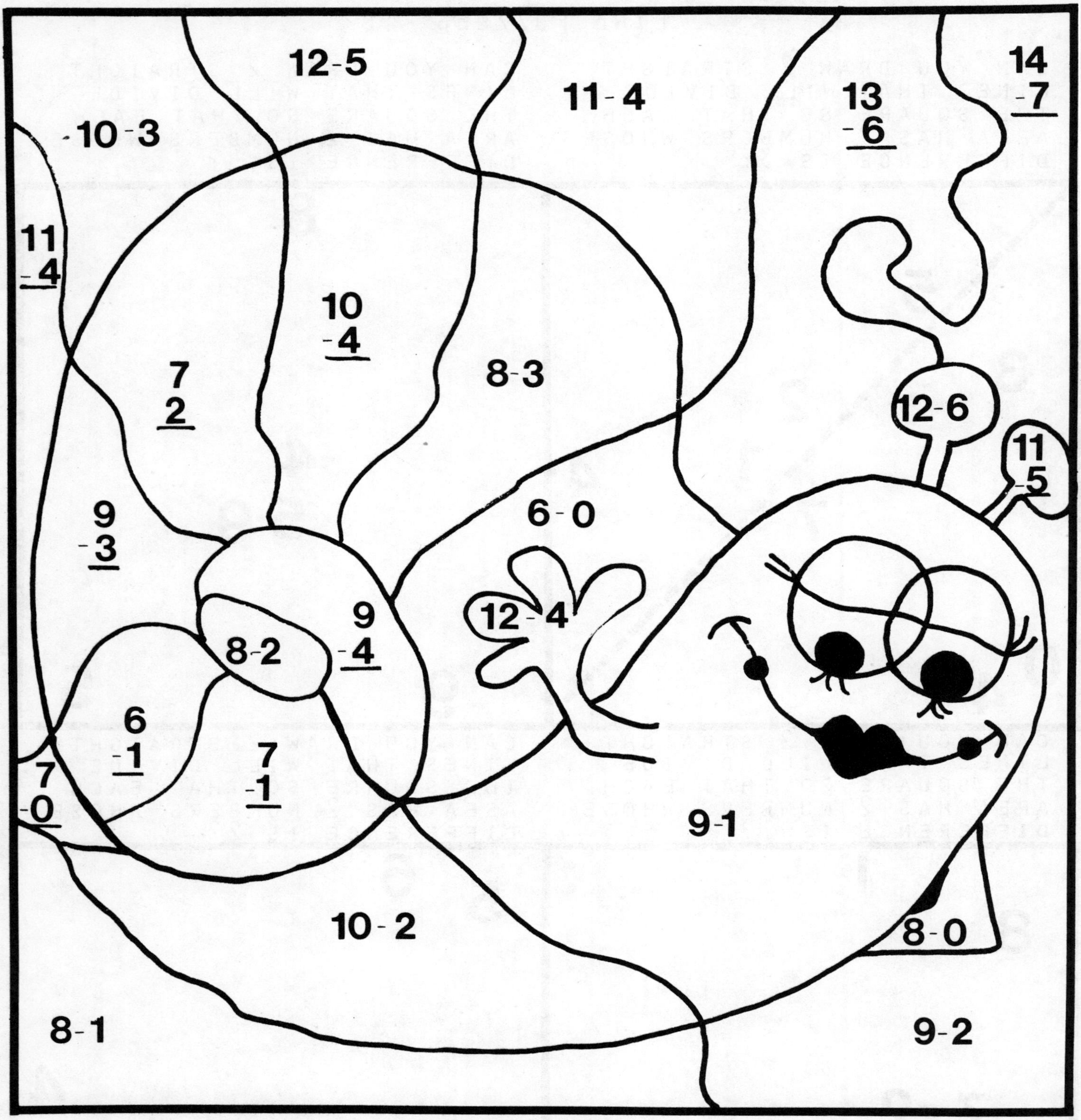

COLOR THE DIFFERENCES OF FIVE (5) RED.

COLOR THE DIFFERENCES OF SIX (6) ORANGE.

COLOR THE DIFFERENCES OF SEVEN (7) BROWN.

COLOR THE DIFFERENCES OF EIGHT (8) YELLOW.

CIRCLE THE NUMBER PAIRS WITH A DIFFERENCE OF 5, PINK. CIRCLE THE NUMBER PAIRS WITH A DIFFERENCE OF 6, ORANGE.

* THE LARGE NUMBER MUST COME FIRST.

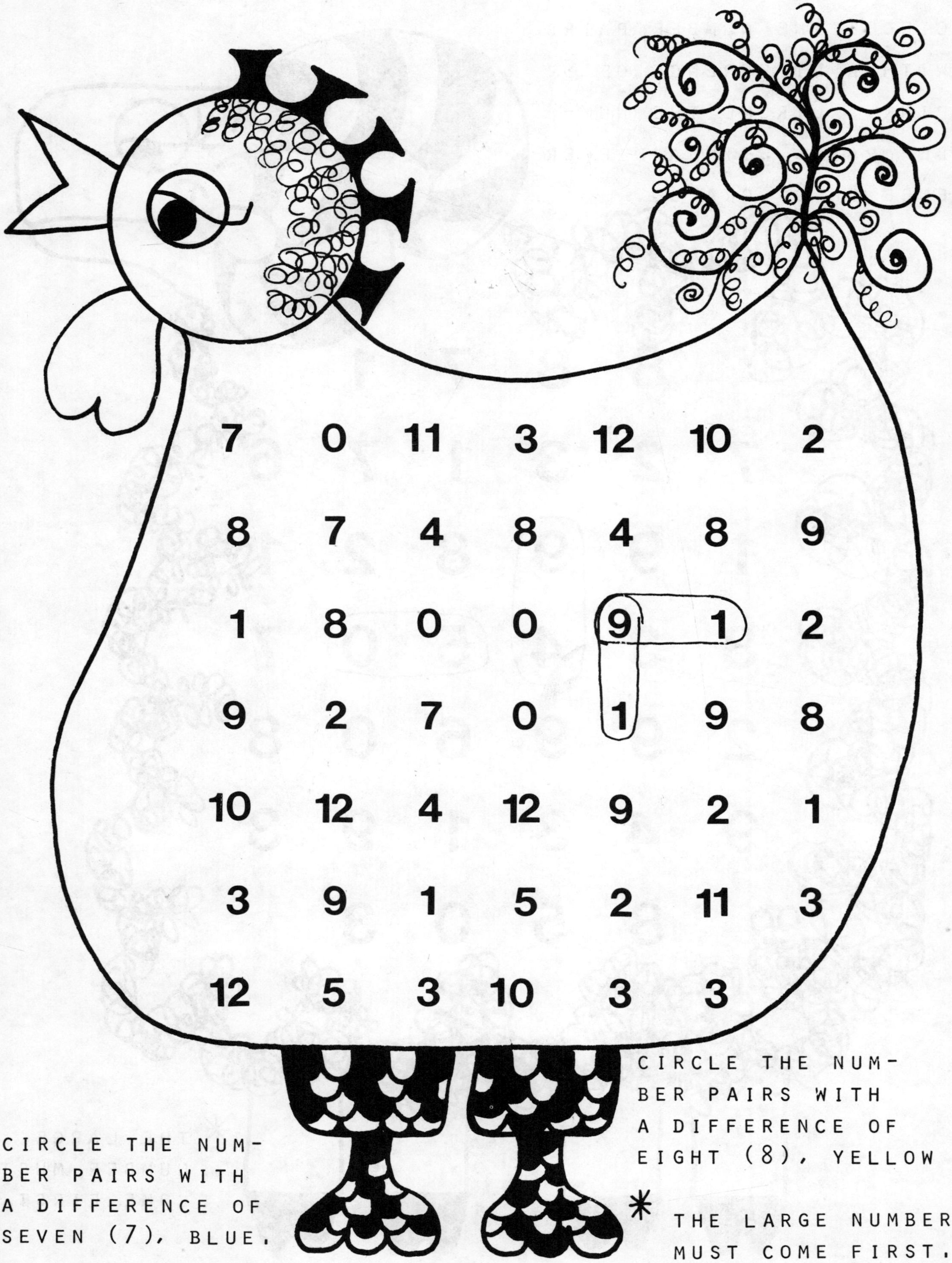

7	0	11	3	12	10	2
8	7	4	8	4	8	9
1	8	0	0	9	1	2
9	2	7	0	1	9	8
10	12	4	12	9	2	1
3	9	1	5	2	11	3
12	5	3	10	3	3	

CIRCLE THE NUMBER PAIRS WITH A DIFFERENCE OF SEVEN (7), BLUE.

CIRCLE THE NUMBER PAIRS WITH A DIFFERENCE OF EIGHT (8), YELLOW

* THE LARGE NUMBER MUST COME FIRST.

DIRECTIONS: WITH SCISSORS CUT APART THE EQUATIONS AT THE BOTTOM OF THE PAGE. GLUE EACH EQUATION IN THE CORRECT BOX.

5	6
7	8

9 - 4	8 - 3	8 - 1
8 - 2	7 - 2	10 - 2
6 - 0	9 - 2	7 - 0
9 - 1	8 - 0	11 - 3
6 - 1	5 - 0	7 - 1
9 - 3	10 - 4	10 - 3

MAGIC SQUARES

9		
4	2	
		3

7	3	
	2	
5		

DIRECTIONS: THE DIFFERENCE BETWEEN THE FIRST TWO NUMBERS SHOULD EQUAL THE FIRST TWO NUMBERS GOING ACROSS AND DOWN

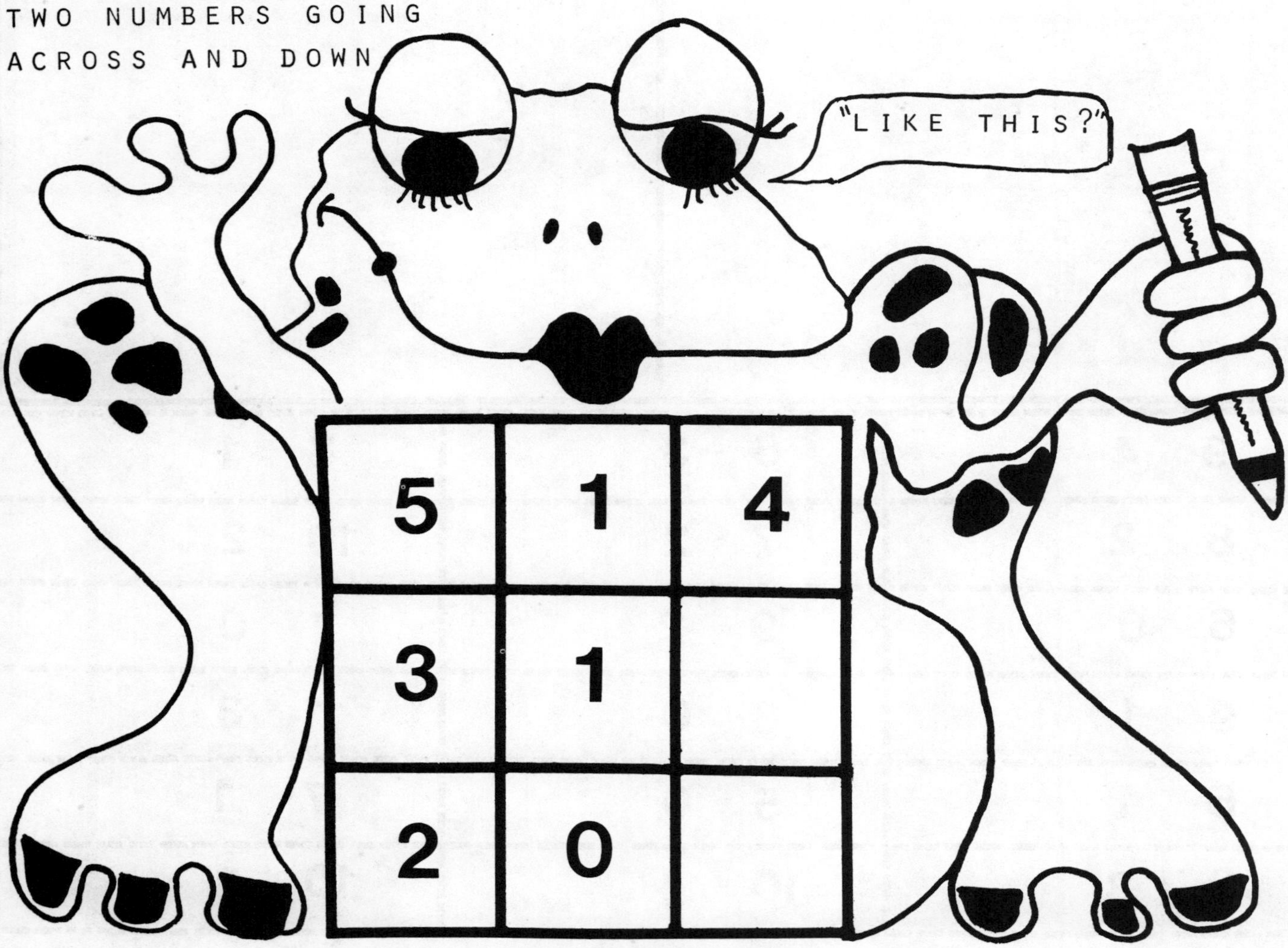

5	1	4
3	1	
2	0	

MAGIC SQUARES

9	5	
		1
6		

7		6
5		5

4	3	
	1	
		0

2		
0	0	
		1

5		3
	1	
		1

5		4
	0	
3		

MAGIC SQUARES

7		5
	2	
	0	

4		
	0	3
	1	

6		
	5	
0		0

5		
	1	
3	0	

7	1	
		5
	0	

1	0	
	0	
		0

HIDDEN CIRCLES

COLOR THE DIFFERENCES OF FOUR (4), RED.

COLOR THE DIFFERENCES OF FIVE (5), PINK.

COLOR THE DIFFERENCES OF SIX (6), YELLOW.

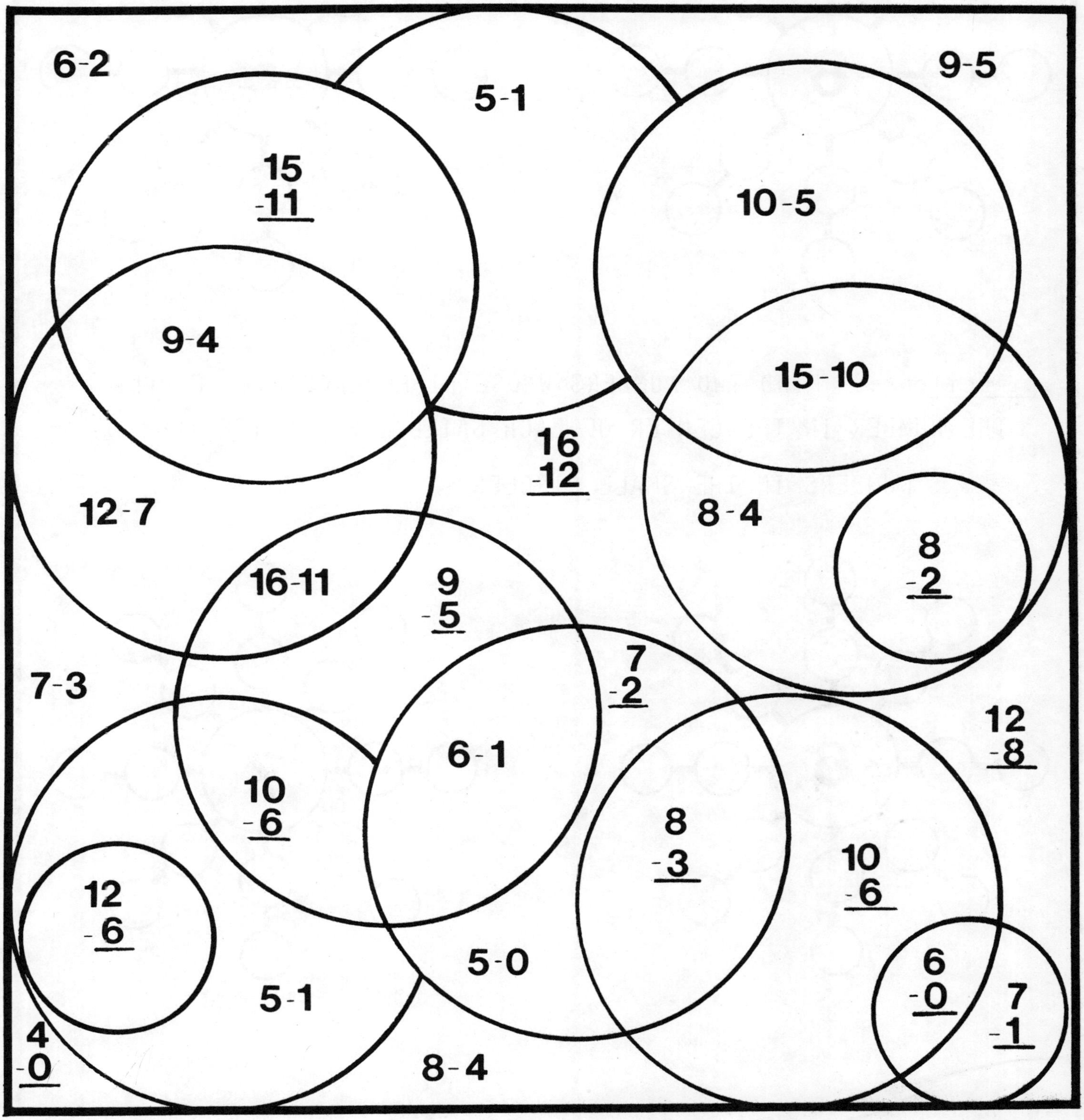

SATELLITE SUBTRACTION

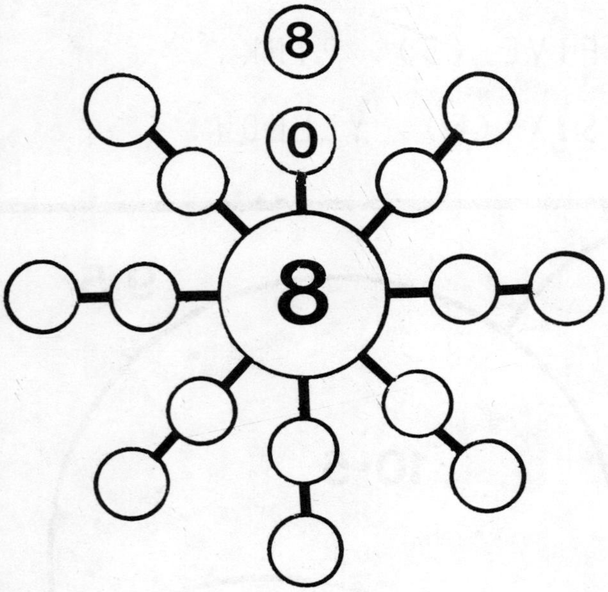

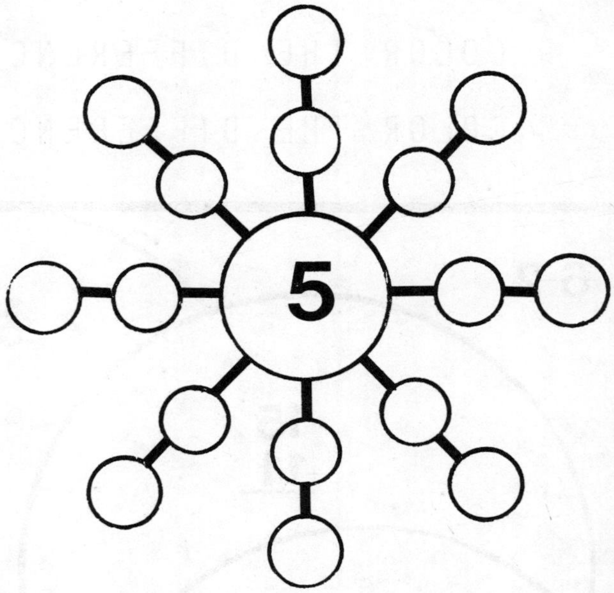

DIRECTIONS: FIND TWO NUMBERS WHOSE DIFFERENCE WILL EQUAL THE NUMBER IN THE CENTER OF EACH SATELLITE. WRITE THE NUMBERS IN THE SMALL CIRCLES.

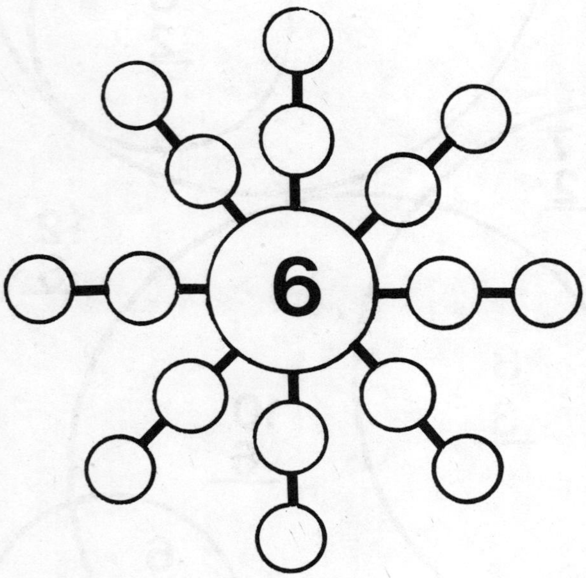

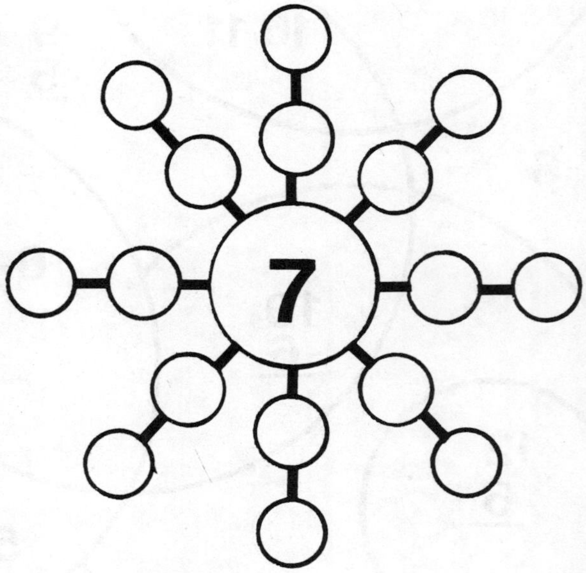

SUBTRACTION WHEELS

DIRECTIONS: SUBTRACT THE NUMBER IN THE CENTER OF EACH WHEEL FROM EACH NUMBER IN THE INNER RING OF EACH WHEEL. WRITE YOUR ANSWER, THE DIFFERENCE, IN EACH SPACE ON THE OUTSIDE RING OF EACH WHEEL.

5 - 4	4 - 1	6 - 5	2 - 0
7 - 5	3 - 2	6 - 4	5 - 2
9 - 6	4 - 2	3 - 0	8 - 7
8 - 4	12 - 7	9 - 5	6 - 1
10 - 5	10 - 4	5 - 0	9 - 2
12 - 6	10 - 6	7 - 0	5 - 1
8 - 1	9 - 2	9 - 1	8 - 0

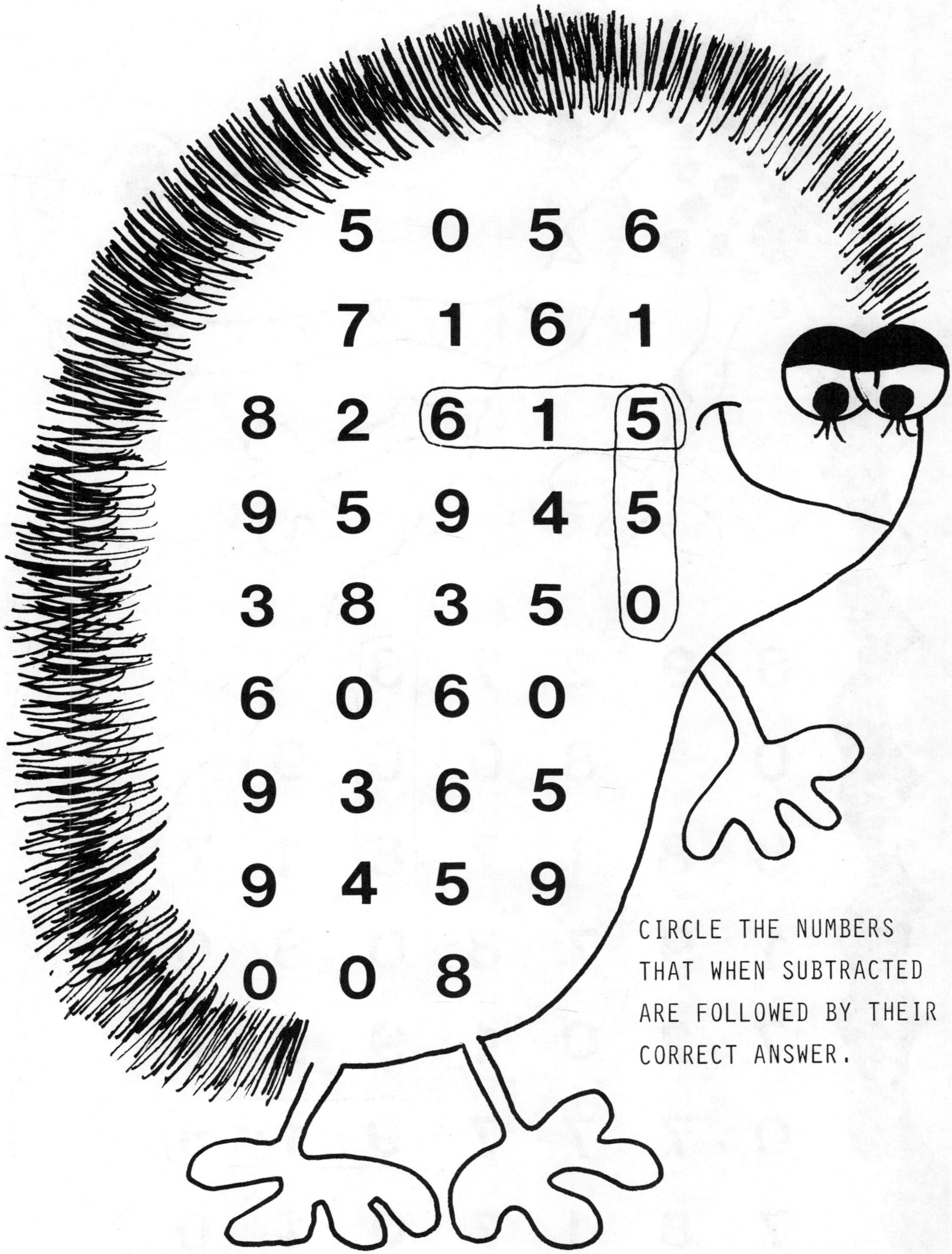

CIRCLE THE NUMBERS THAT WHEN SUBTRACTED ARE FOLLOWED BY THEIR CORRECT ANSWER.

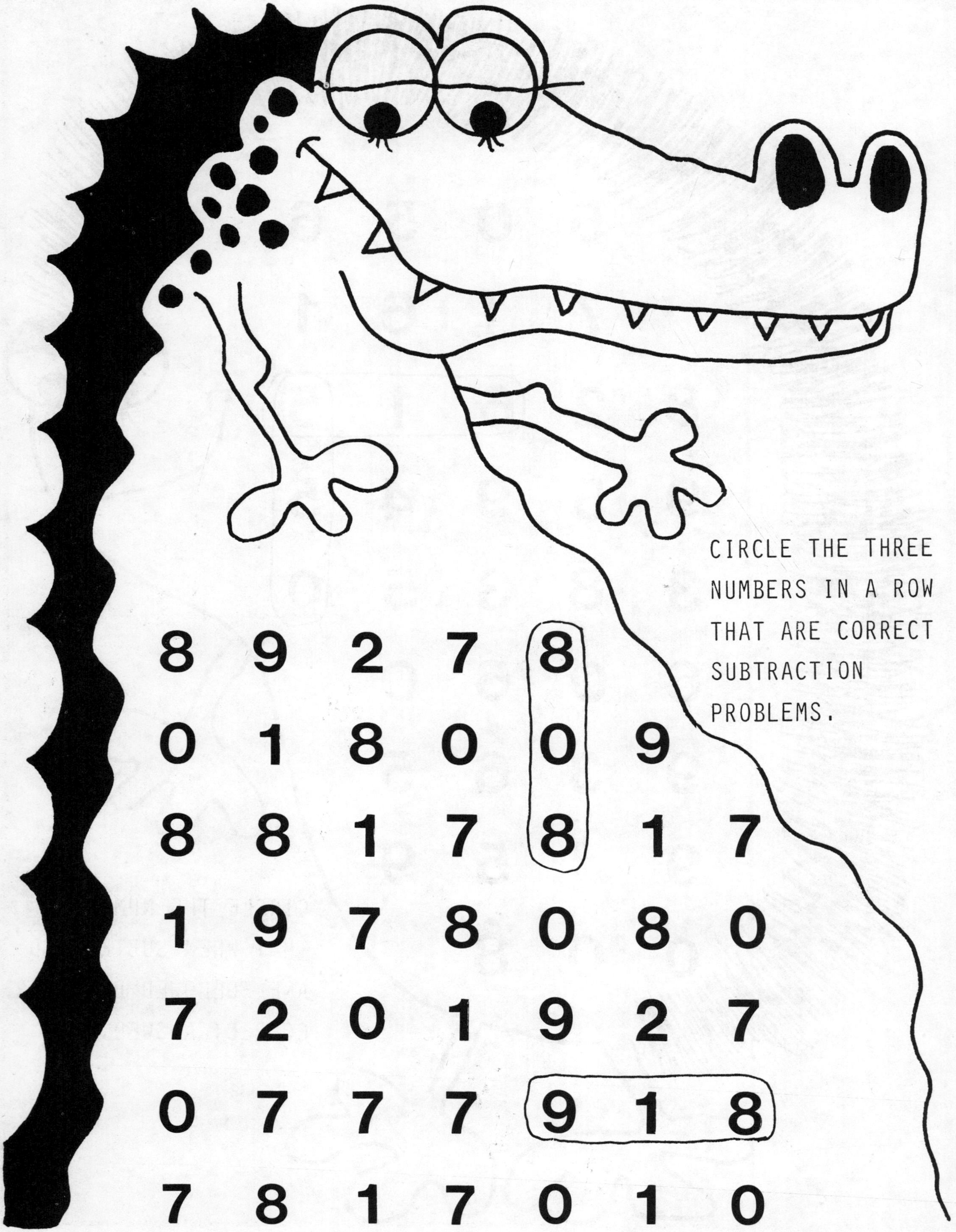

CIRCLE THE THREE NUMBERS IN A ROW THAT ARE CORRECT SUBTRACTION PROBLEMS.

8	9	2	7	8		
0	1	8	0	0	9	
8	8	1	7	8	1	7
1	9	7	8	0	8	0
7	2	0	1	9	2	7
0	7	7	7	9	1	8
7	8	1	7	0	1	0

DRAW A LINE TO CONNECT THE NUMBERS WITH A DIFFERENCE OF FIVE (5).

5	2
6	0
7	1
8	6
9	3
10	4
11	5

DRAW A LINE TO CONNECT THE NUMBERS WITH A DIFFERENCE OF SIX (6).

6	5
7	0
8	4
9	3
10	6
11	2
12	1

DRAW A LINE TO CONNECT THE NUMBERS WITH A DIFFERENCE OF SEVEN (7).

7	2
8	0
9	1
10	6
11	3
12	4
13	5

DRAW A LINE TO CONNECT THE NUMBERS WITH A DIFFERENCE OF EIGHT (8).

8	1
9	0
10	2
11	6
12	5
13	4
14	3

DIRECTIONS: START IN THE CIRCLES MARKED WITH THE NUMBER FIFTEEN (15). MOVE ALONG THE PATH CONNECTING THE CIRCLES. SUBTRACT EACH NUMBER AS YOU PASS THROUGH THE CIRCLE. WRITE THE DIFFERENCE IN THE EMPTY CIRCLE AT THE BOTTOM OF THE PAGE.

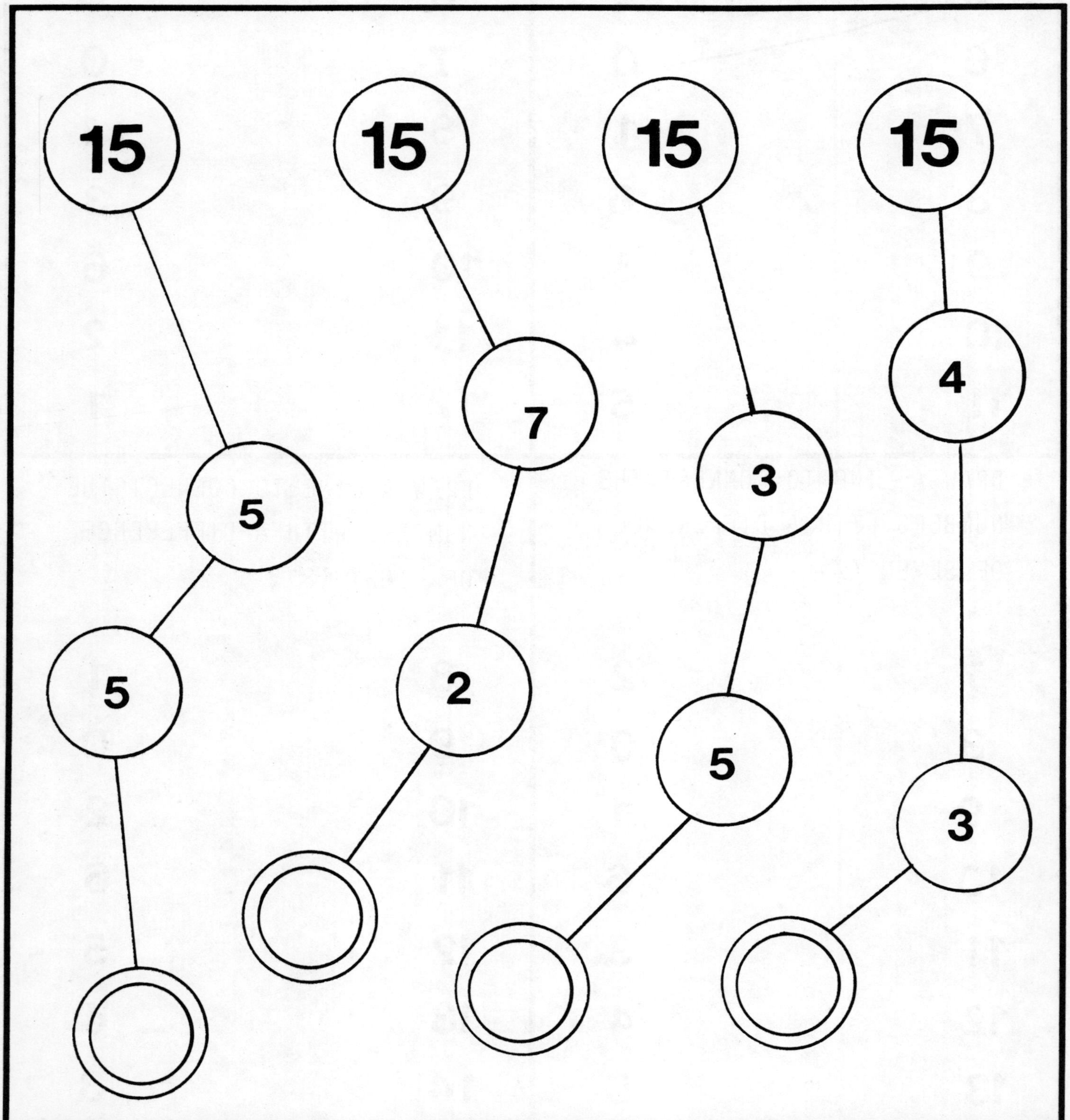

BEGIN AT THE "GO". YOU HAVE 25 POINTS. AS YOU TRAVEL THROUGH A CIRCLE, SUBTRACT "THAT MANY POINTS". CAN YOU FIND A TRAIL:

THAT WILL GIVE YOU EXACTLY 5 POINTS WHEN YOU ARE AT "END"?

THAT WILL GIVE YOU EXACTLY 6 POINTS WHEN YOU ARE AT "END"?

THAT WILL GIVE YOU EXACTLY 7 POINTS WHEN YOU ARE AT "END"?

THAT WILL GIVE YOU EXACTLY 8 POINTS WHEN YOU ARE AT "END"?

CAN YOU FIND SOME MORE TRAILS TO FOLLOW? WHAT IS THE NUMBER OF POINTS YOU HAVE LEFT WHEN YOU REACH END ON THESE TRAILS?

TRAIL FOLLOWED	ANSWER
______________________________	____________
______________________________	____________

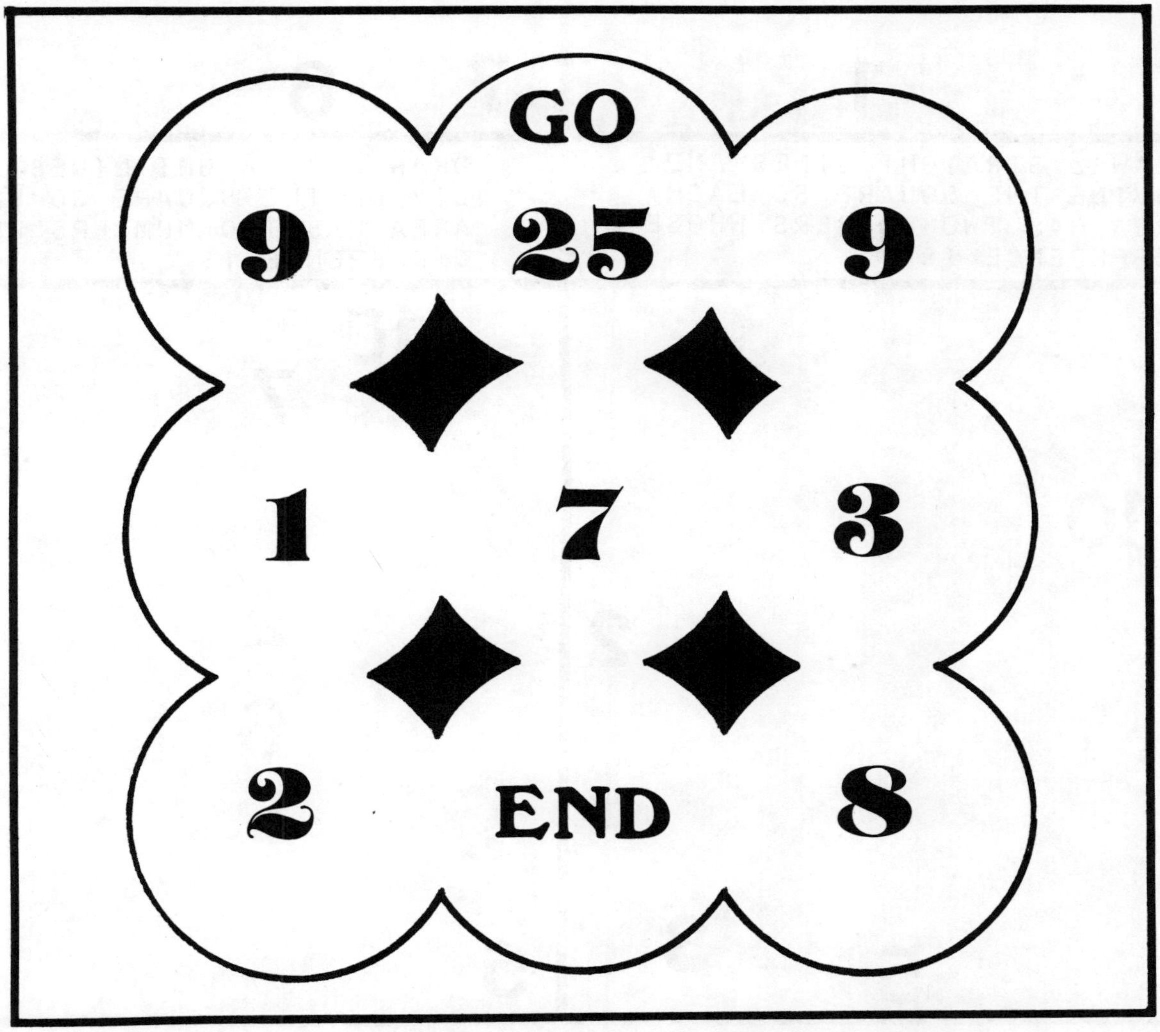

LINE PUZZLES

CAN YOU DRAW 2 STRAIGHT LINES THAT WILL DIVIDE THE SQUARE SO THAT EACH AREA HAS 2 NUMBERS WHOSE DIFFERENCE IS 6.

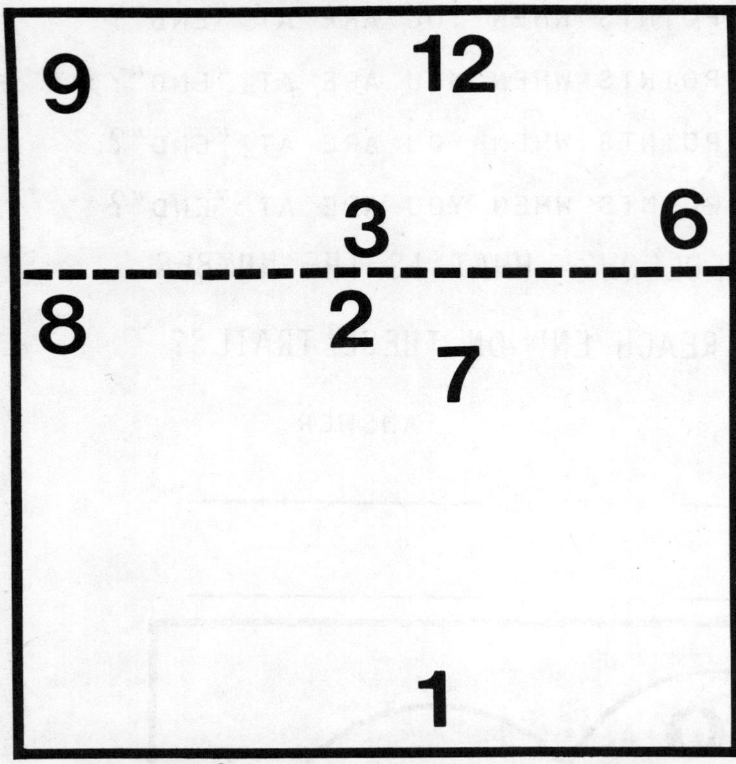

CAN YOU DRAW 2 STRAIGHT LINES THAT WILL DIVIDE THE SQUARE SO THAT EACH AREA HAS 2 NUMBERS WHOSE DIFFERENCE IS 0.

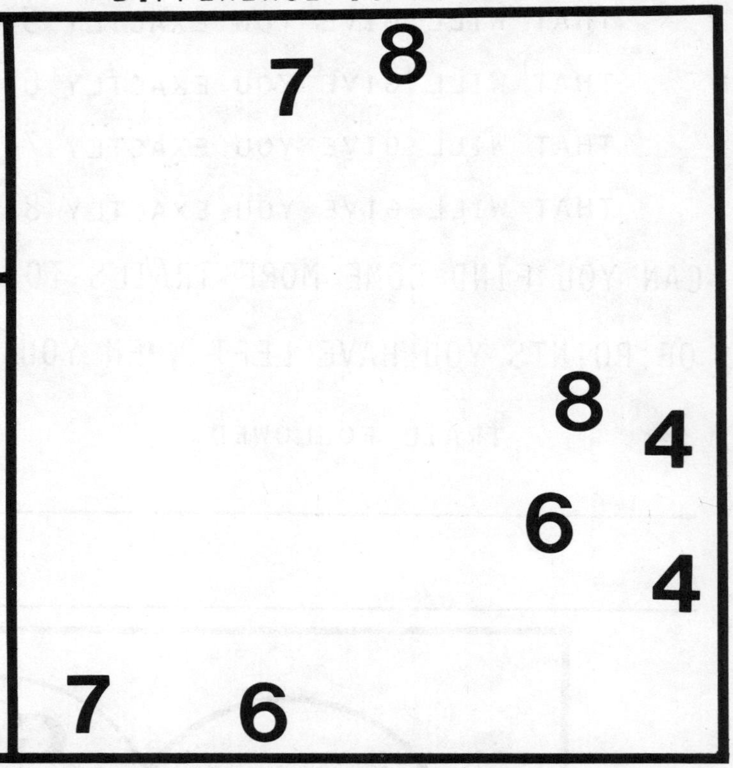

DRAW 2 STRAIGHT LINES AND DIVIDE THE SQUARE SO EACH AREA HAS TWO NUMBERS WHOSE DIFFERENCE IS 7

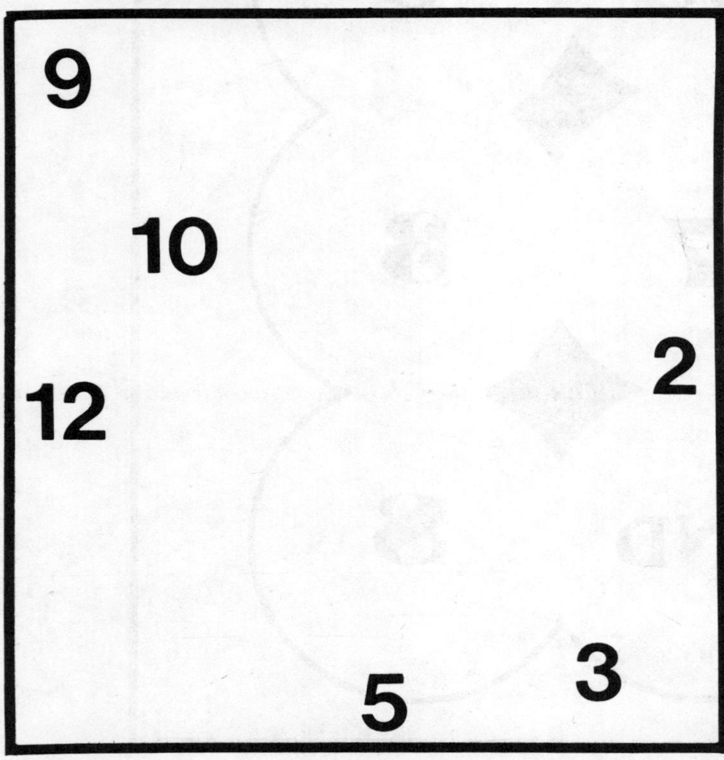

DRAW 2 STRAIGHT LINES AND DIVIDE THE SQUARE SO EACH AREA HAS TWO NUMBERS WHOSE DIFFERENCE IS 5.

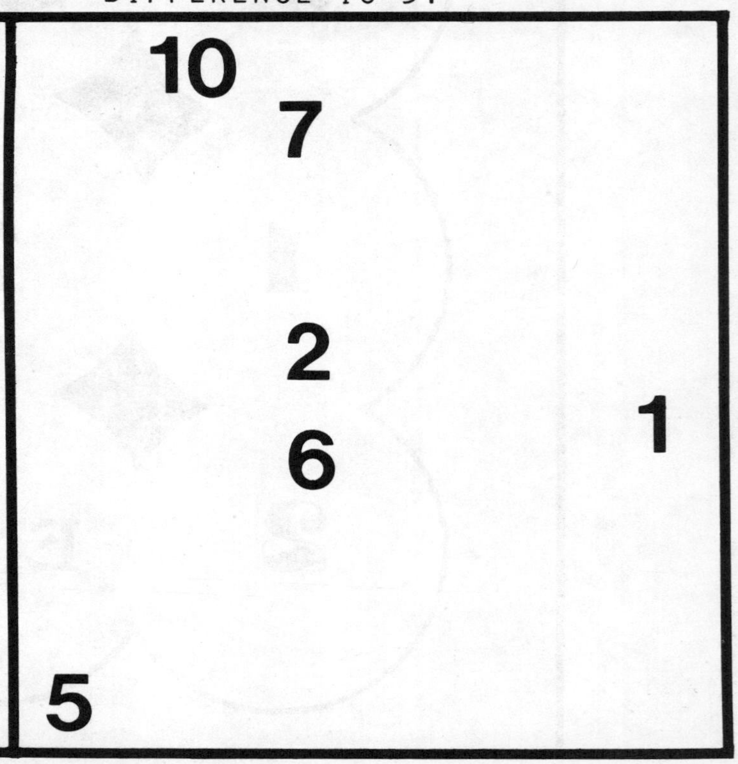

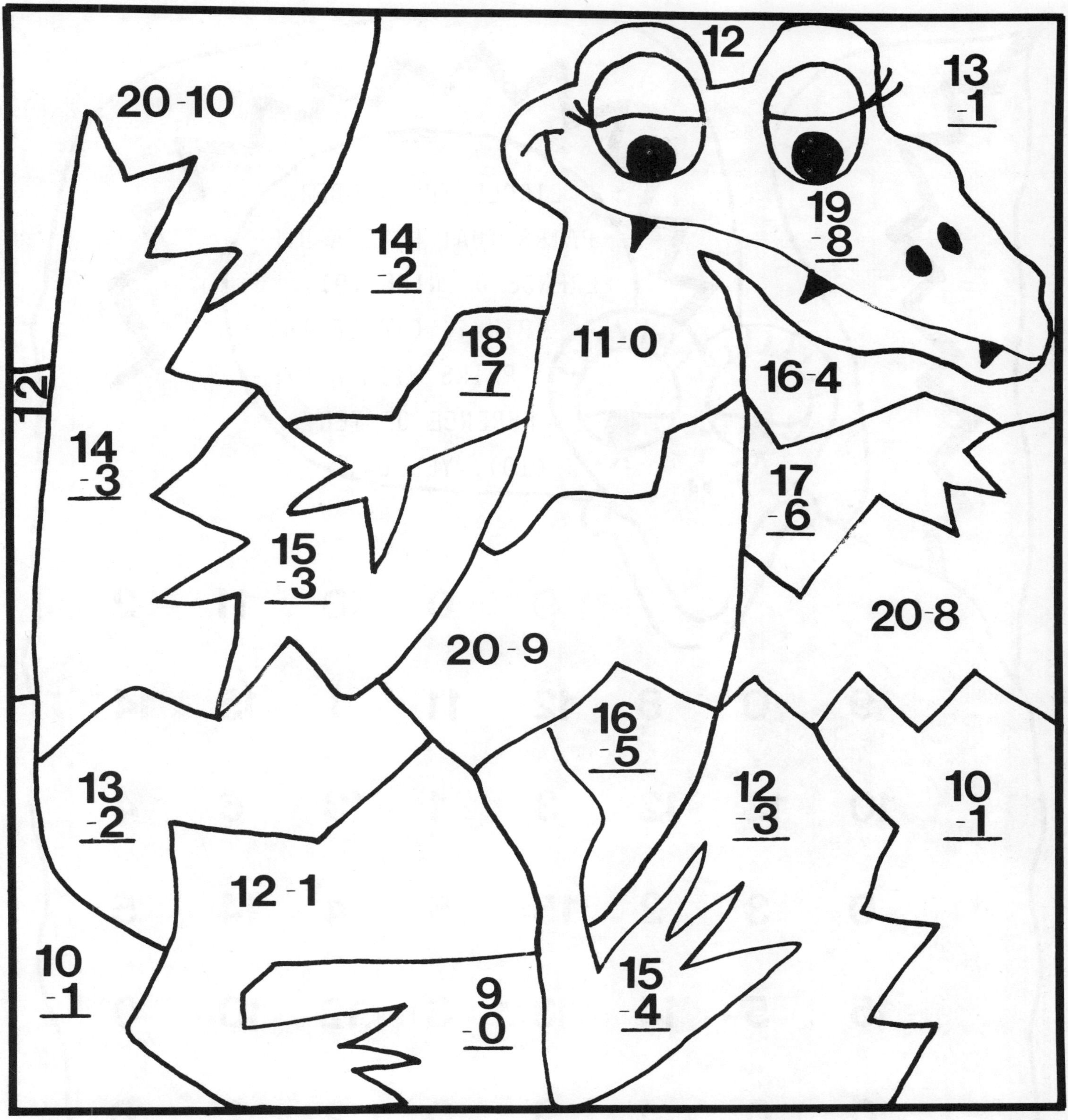

DIRECTIONS:

IF THE DIFFERENCE IS NINE (9), COLOR THE AREA GREEN.

IF THE DIFFERENCE IS TEN (10), COLOR THE AREA ORANGE.

IF THE DIFFERENCE IS ELEVEN (11), COLOR THE AREA PURPLE.

IF THE DIFFERENCE IS TWELVE (12), COLOR THE AREA YELLOW.

"CIRCLE THE NUMBER PAIRS THAT HAVE A DIFFERENCE OF NINE (9), PINK. CIRCLE THOSE PAIRS WITH A DIFFERENCE OF TEN (10), YELLOW."
9 0 10 11 2
9 0 8 12 11 1 15 14
10 13 12 3 1 13 6 4
0 3 2 15 5 4 14 5
15 5 14 13 3 12 10 0
10 0 4 9 0 2 1 6
1 15 6 0 14 5 6

CIRCLE THE NUMBER PAIRS WITH A DIFFERENCE OF ELEVEN (11), GREEN. CIRCLE THE NUMBER PAIRS WITH A DIFFERENCE OF TWELVE (12), ORANGE.

14	3	17	5	19	8	7	8
12	1	18	19	7	20	18	19
16	12	6	13	14	9	7	8
5	0	9	1	2	13	2	14
17	6	15	15	3	18	19	3
16	17	4	12	1	7	8	12
4	5	7	17	6	13	2	0

DIRECTIONS: WITH SCISSORS CUT APART THE EQUATIONS AT THE BOTTOM OF THE PAGE. GLUE EACH EQUATION IN THE CORRECT BOX AT THE TOP OF THE PAGE.

9	10
11	12

9 - 0	12 - 1	11 - 1
12 - 2	10 - 0	13 - 2
14 - 2	10 - 1	11 - 0
13 - 3	11 - 2	12 - 3
14 - 4	15 - 3	14 - 3
12 - 0	13 - 1	13 - 4

HIDDEN CIRCLES

COLOR THE AREAS WITH A DIFFERENCE OF SEVEN (7), RED.
COLOR THE AREAS WITH A DIFFERENCE OF EIGHT (8), YELLOW
COLOR THE AREAS WITH A DIFFERENCE OF NINE (9), ORANGE.

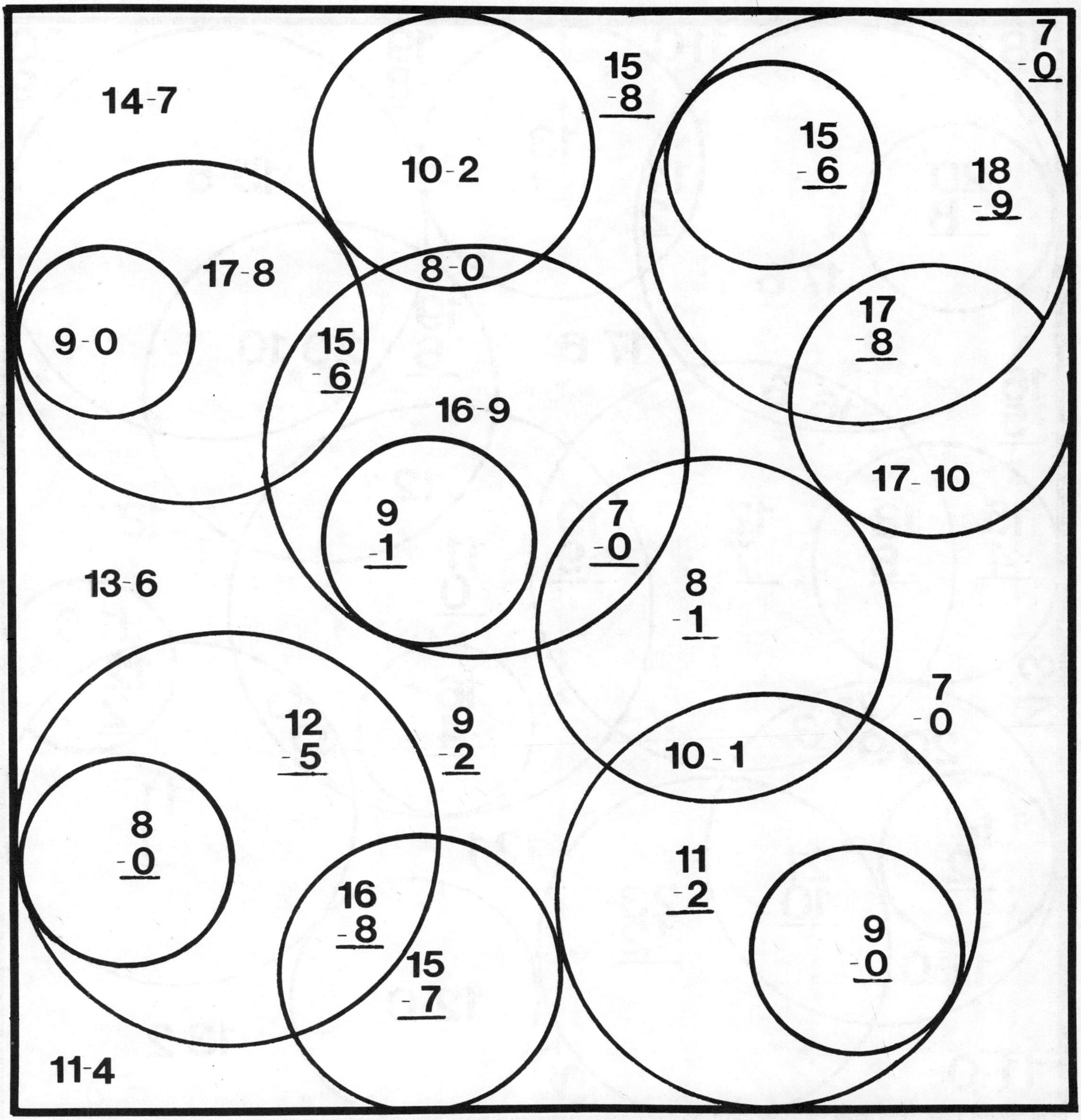

HIDDEN CIRCLES

COLOR THE AREAS WITH A DIFFERENCE OF TEN (10), BROWN.

COLOR THE AREAS WITH A DIFFERENCE OF ELEVEN (11), YELLOW.

COLOR THE AREAS WITH A DIFFERENCE OF TWELVE (12), PINK.

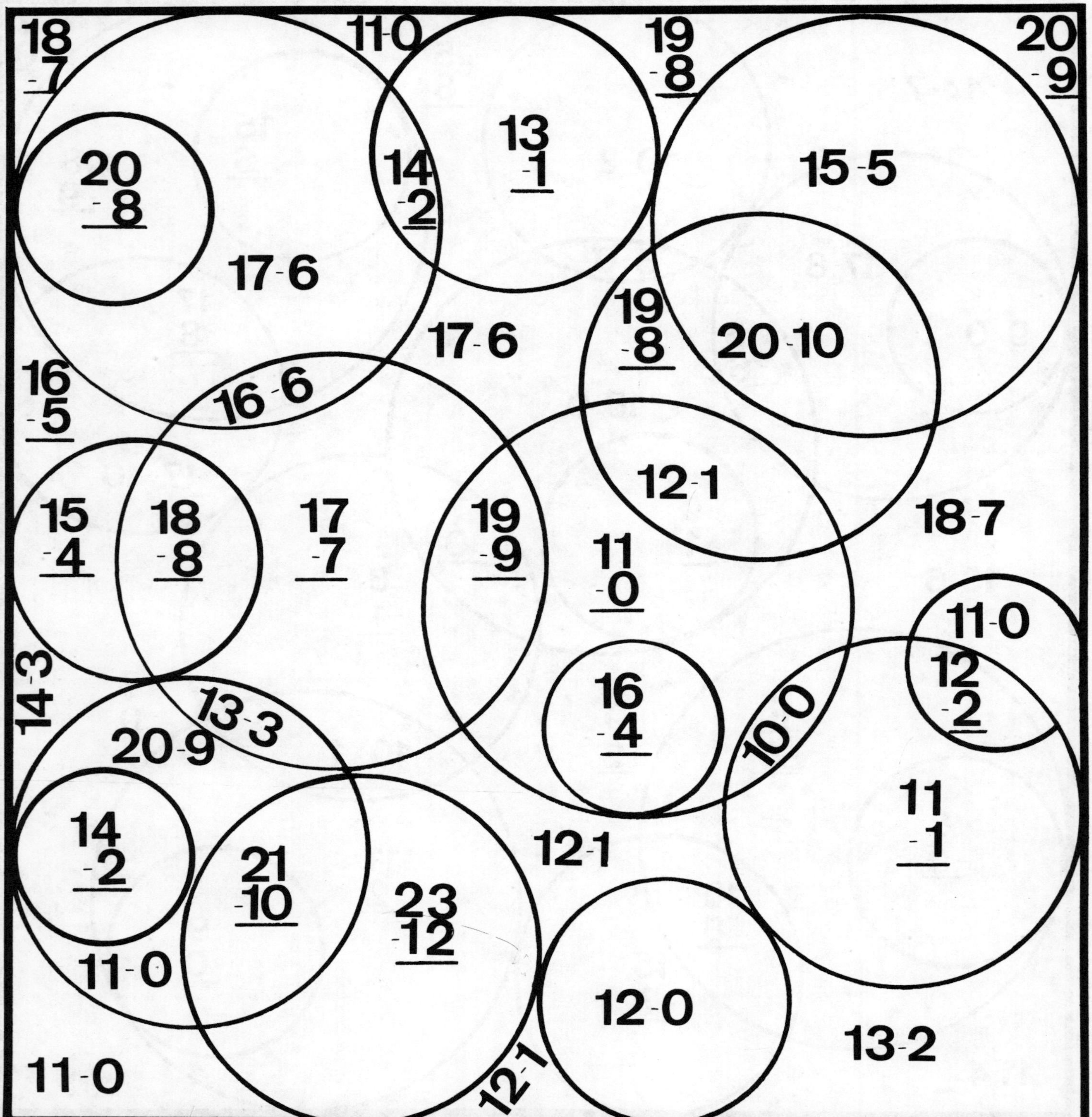

	11	12	13	14	15	16	17	18	19
-1									
-2									
-3									
-4									
-5				9					
-6									
-7									
-8					7				
-9									
-10			3						
-11									

COMPLETE THIS SUBTRACTION CHART. DO YOU SEE ANY NUMBER PATTERNS? COLOR ODD NUMBERED SQUARES, YELLOW. COLOR EVEN NUMBERED SQUARES, ORANGE. NOW YOU CAN SEE THE PATTERN.

10 - 5	14 - 7
12 - 6	11 - 6
12 - 5	10 - 4
16 - 8	9 - 0
20 - 10	10 - 2
10 - 1	11 - 1

15 - 5	12 - 2
18 - 9	12 - 3
16 - 8	8 - 1
10 - 3	9 - 1
10 - 4	8 - 3
7 - 2	9 - 3

DRAW A LINE TO CONNECT THE SUBTRACTION PROBLEMS THAT HAVE THE SAME DIFFERENCE.

20 - 9	20 - 7
25 - 13	21 - 10
25 - 12	20 - 8
20 - 14	12 - 5
19 - 12	12 - 6
15 - 7	20 - 11
10 - 1	14 - 6
14 - 4	14 - 14
15 - 15	16 - 6
21 - 6	25 - 10
21 - 7	7 - 4
9 - 6	20 - 6

SATELLITE SUBTRACTION

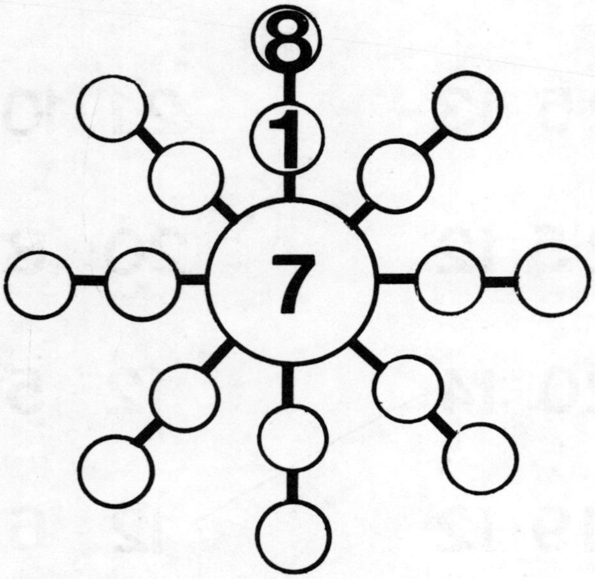

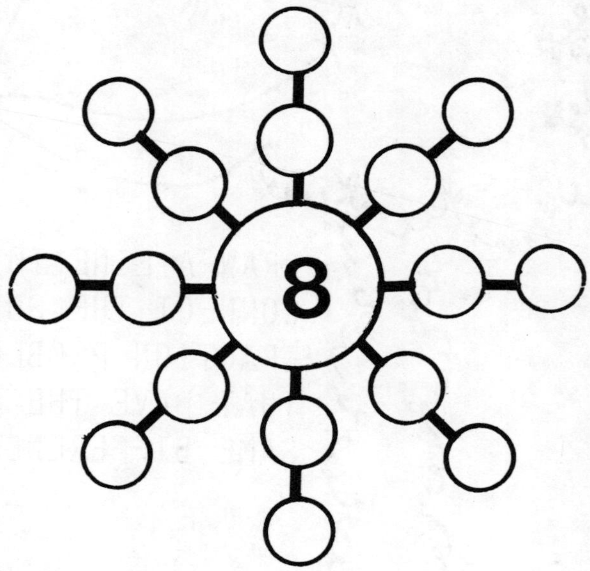

<u>DIRECTIONS:</u>

FIND TWO NUMBERS WHOSE DIFFERENCE EQUALS THE NUMBER IN THE CENTER OF EACH SATELLITE. WRITE IN ALL THE MISSING NUMBERS.

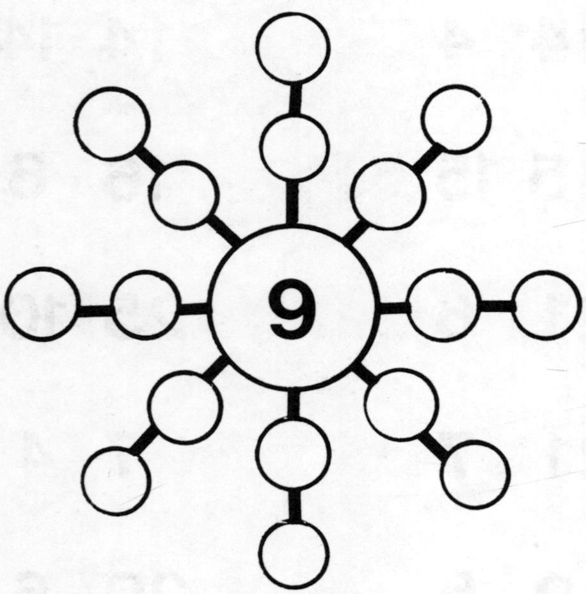

SUBTRACTION WHEELS

<u>DIRECTIONS:</u> SUBTRACT THE NUMBER IN THE CENTER OF EACH WHEEL FROM EACH NUMBER IN THE INNER RING OF EACH WHEEL. WRITE YOUR ANSWER, THE DIFFERENCE, IN EACH SPACE ON THE OUTSIDE RING OF EACH WHEEL.

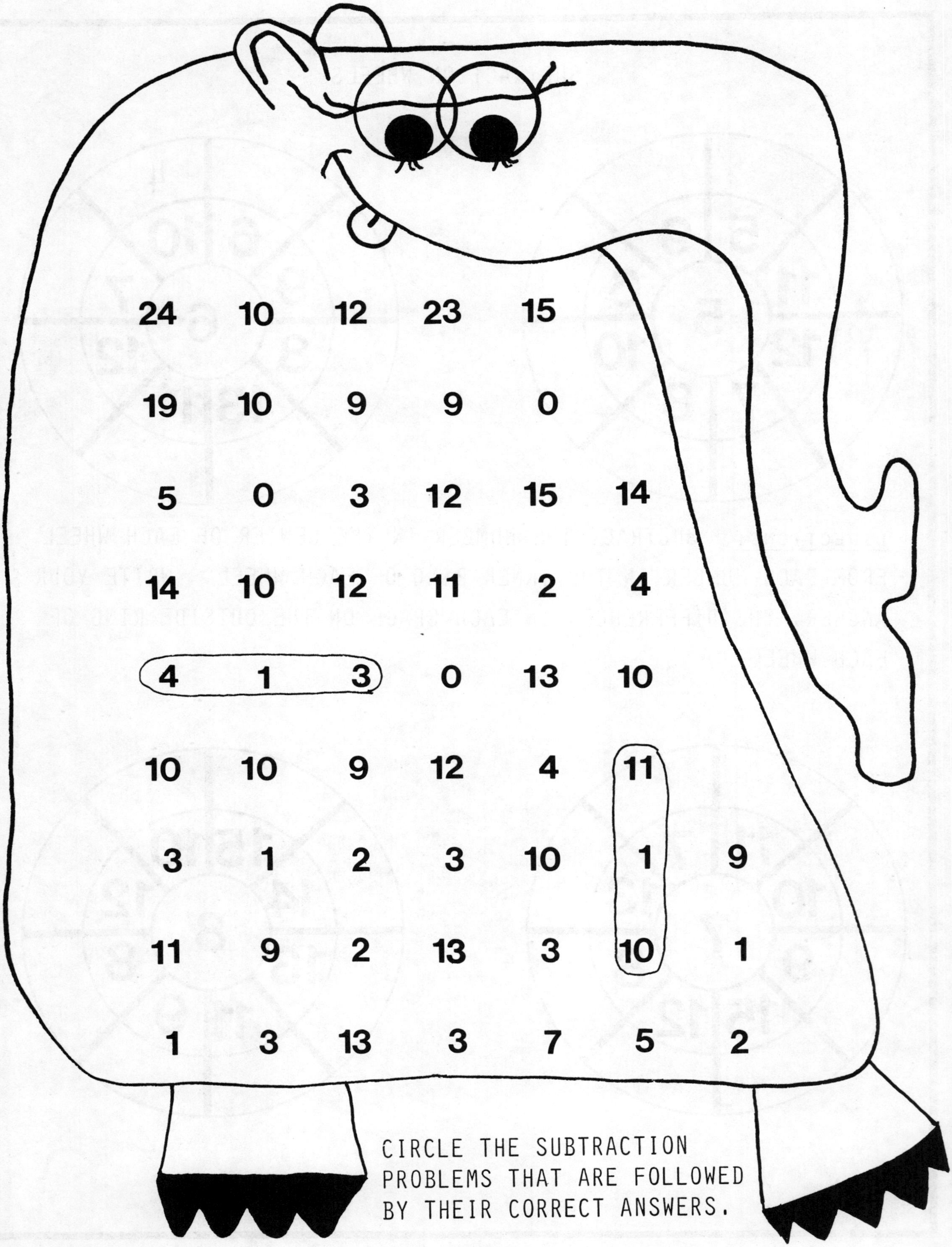

24	10	12	23	15		
19	10	9	9	0		
5	0	3	12	15	14	
14	10	12	11	2	4	
4	1	3	0	13	10	
10	10	9	12	4	11	
3	1	2	3	10	1	9
11	9	2	13	3	10	1
1	3	13	3	7	5	2

CIRCLE THE SUBTRACTION PROBLEMS THAT ARE FOLLOWED BY THEIR CORRECT ANSWERS.

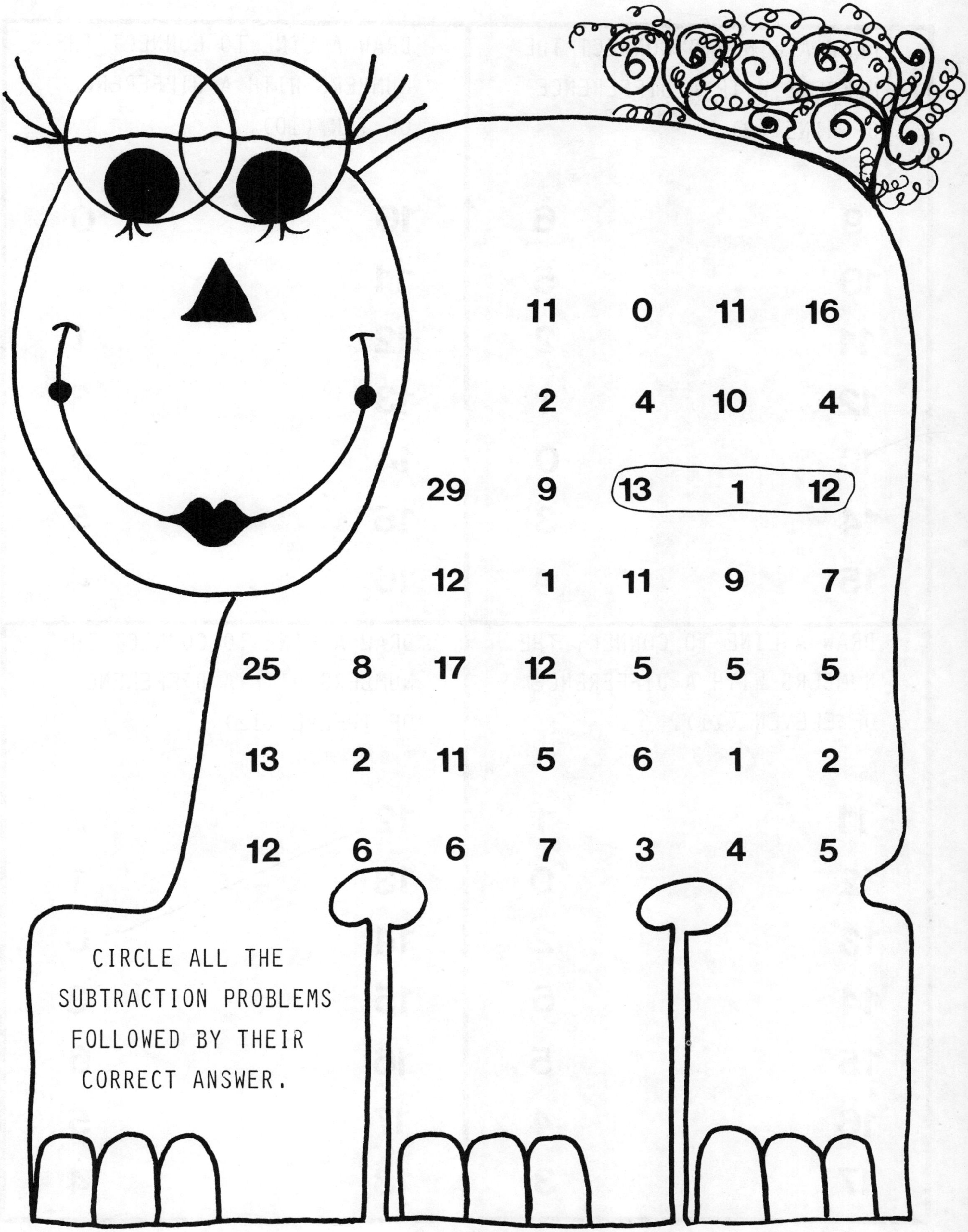
11 0 11 16
2 4 10 4
29 9 13 1 12
12 1 11 9 7
25 8 17 12 5 5 5
13 2 11 5 6 1 2
12 6 6 7 3 4 5
CIRCLE ALL THE SUBTRACTION PROBLEMS FOLLOWED BY THEIR CORRECT ANSWER.

DRAW A LINE TO CONNECT THE NUMBERS WITH A DIFFERENCE OF NINE (9).

9	6
10	5
11	2
12	1
13	0
14	3
15	4

DRAW A LINE TO CONNECT THE NUMBERS WITH A DIFFERENCE OF TEN (10).

10	0
11	2
12	1
13	3
14	6
15	5
16	4

DRAW A LINE TO CONNECT THE NUMBERS WITH A DIFFERENCE OF ELEVEN (11).

11	1
12	0
13	2
14	6
15	5
16	4
17	3

DRAW A LINE TO CONNECT THE NUMBERS WITH A DIFFERENCE OF TWELVE (12).

12	2
13	1
14	0
15	3
16	6
17	5
18	4

DIRECTIONS: START IN THE CIRCLES MARKED WITH THE NUMBER TWENTY (20). MOVE ALONG THE PATH CONNECTING THE CIRCLES. SUBTRACT EACH NUMBER AS YOU PASS THROUGH THE CIRCLE. WRITE THE DIFFERENCE IN THE EMPTY CIRCLE AT THE BOTTOM OF THE PAGE.

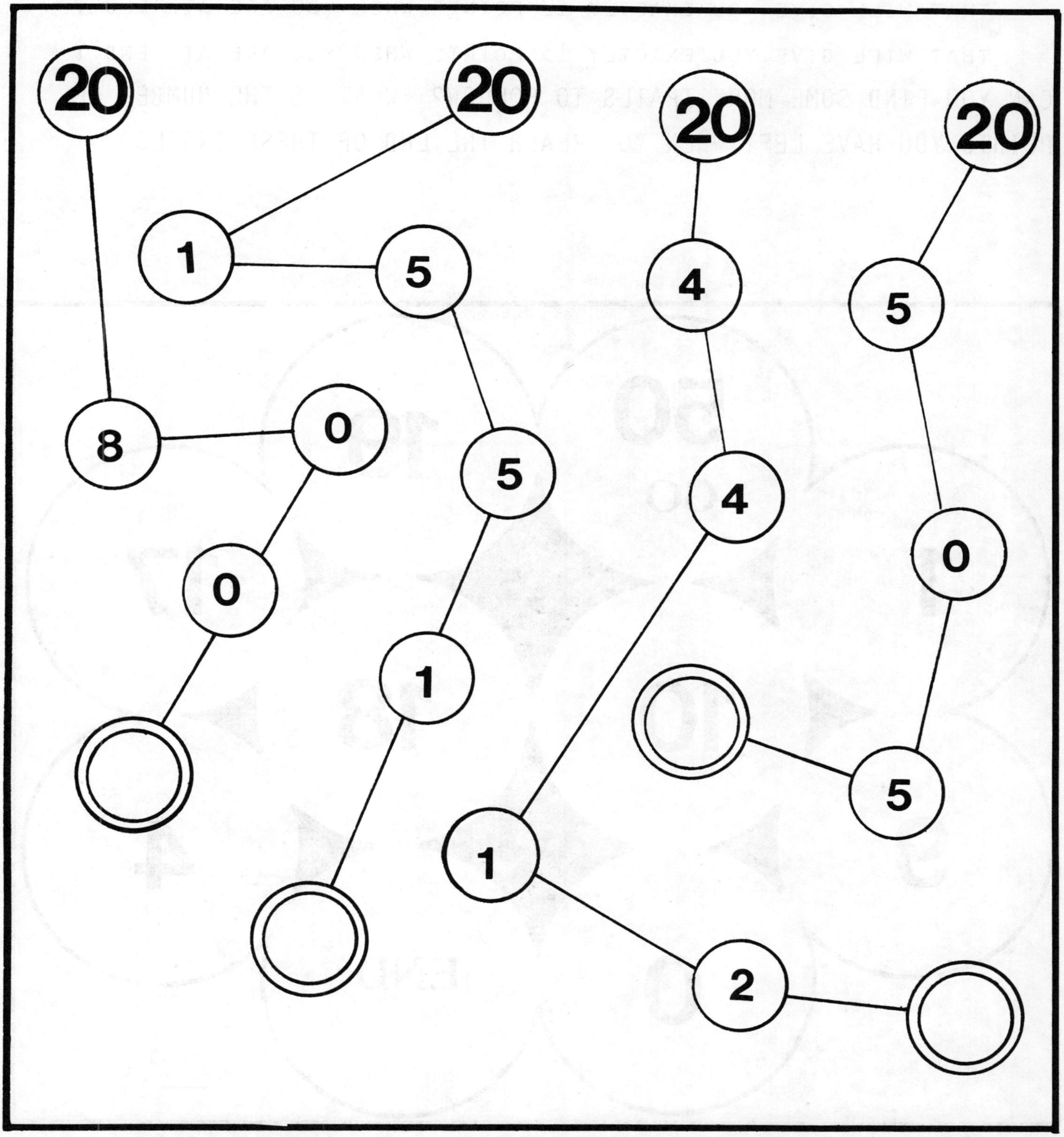

BEGIN AT THE "GO". YOU HAVE 50 POINTS. AS YOU TRAVEL THROUGH A CIRCLE, SUBTRACT "THAT MANY POINTS". CAN YOU FIND A TRAIL:

THAT WILL GIVE YOU EXACTLY 9 POINTS WHEN YOU ARE AT "END"?

THAT WILL GIVE YOU EXACTLY 10 POINTS WHEN YOU ARE AT "END"?

THAT WILL GIVE YOU EXACTLY 12 POINTS WHEN YOU ARE AT "END"?

THAT WILL GIVE YOU EXACTLY 13 POINTS WHEN YOU ARE AT "END"?

CAN YOU FIND SOME MORE TRAILS TO FOLLOW? WHAT IS THE NUMBER OF POINTS YOU HAVE LEFT WHEN YOU REACH THE END OF THESE TRAILS?

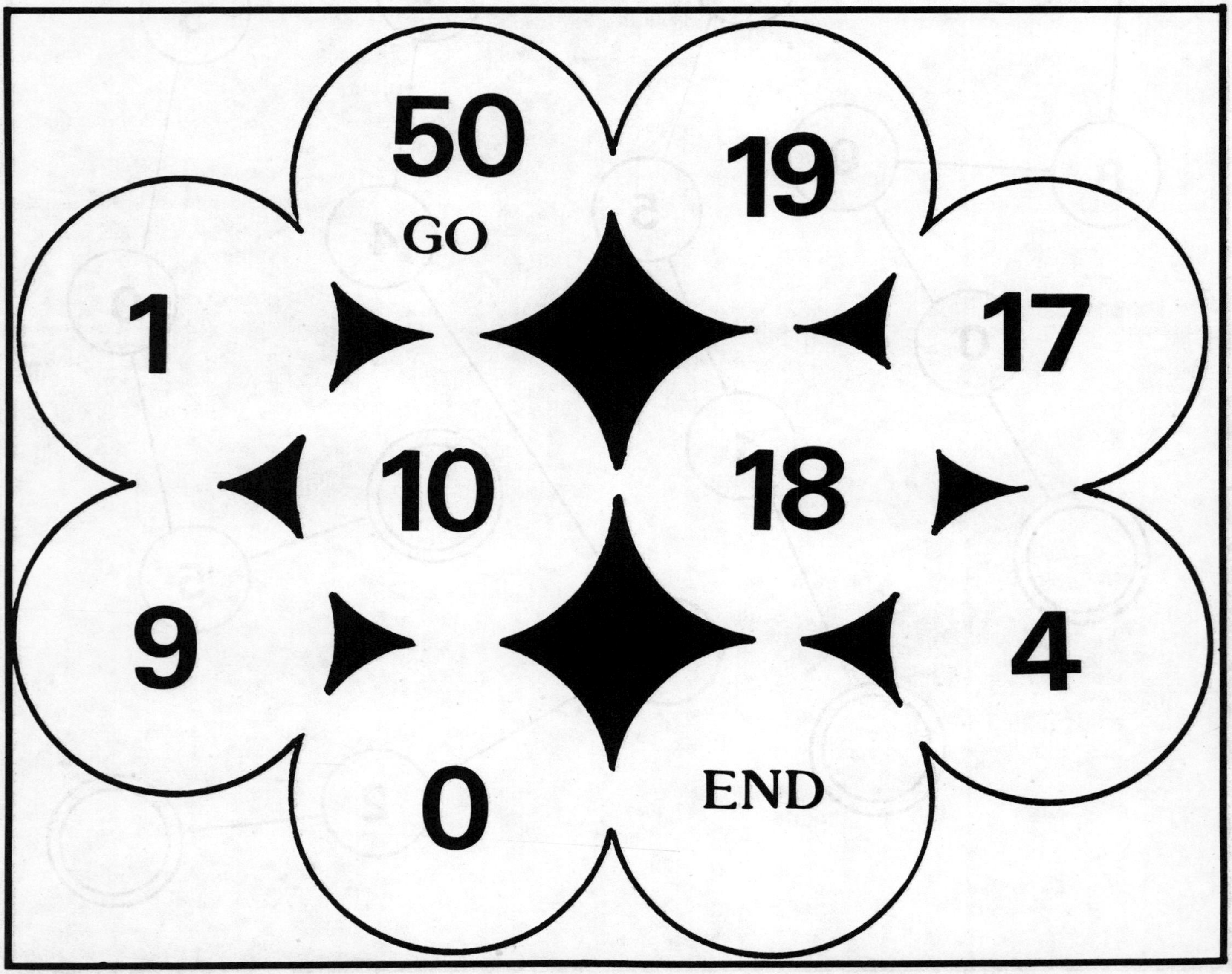

LINE PUZZLES

CAN YOU DRAW 2 STRAIGHT LINES THAT WILL DIVIDE THE SQUARE SO THAT EACH AREA HAS 2 NUMBERS WHOSE DIFFERENCE IS EIGHT (8).

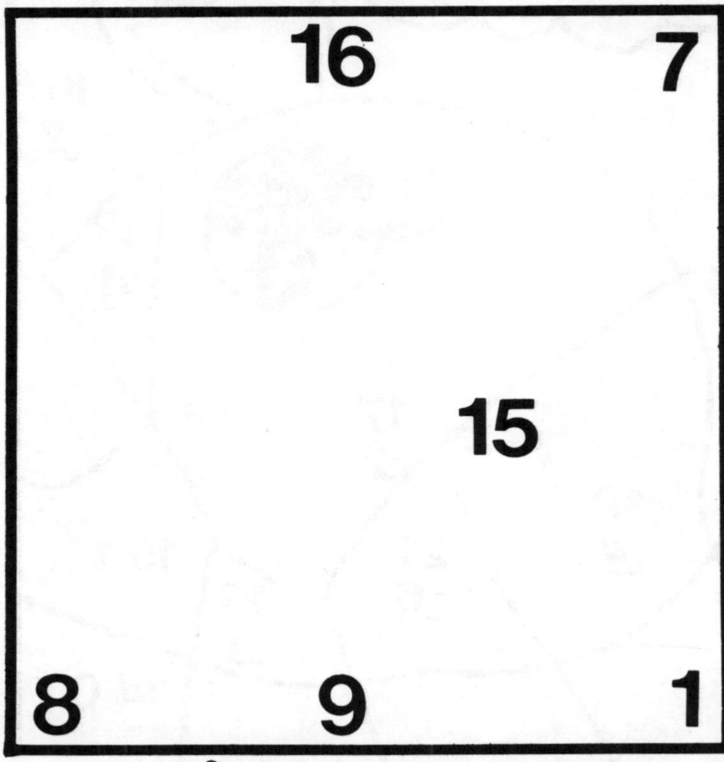

CAN YOU DRAW 2 STRAIGHT LINES THAT WILL DIVIDE THE SQUARE SO THAT EACH AREA HAS 2 NUMBERS WHOSE DIFFERENCE IS NINE (9).

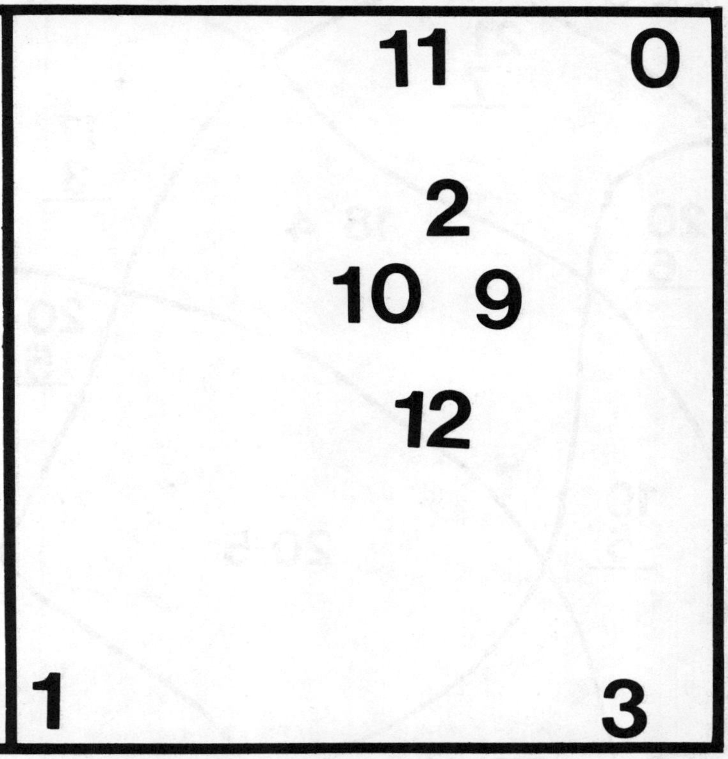

DRAW 2 STRAIGHT LINES AND DIVIDE THE SQUARE SO EACH AREA HAS TWO NUMBERS WHOSE DIFFERENCE IS TEN (10).

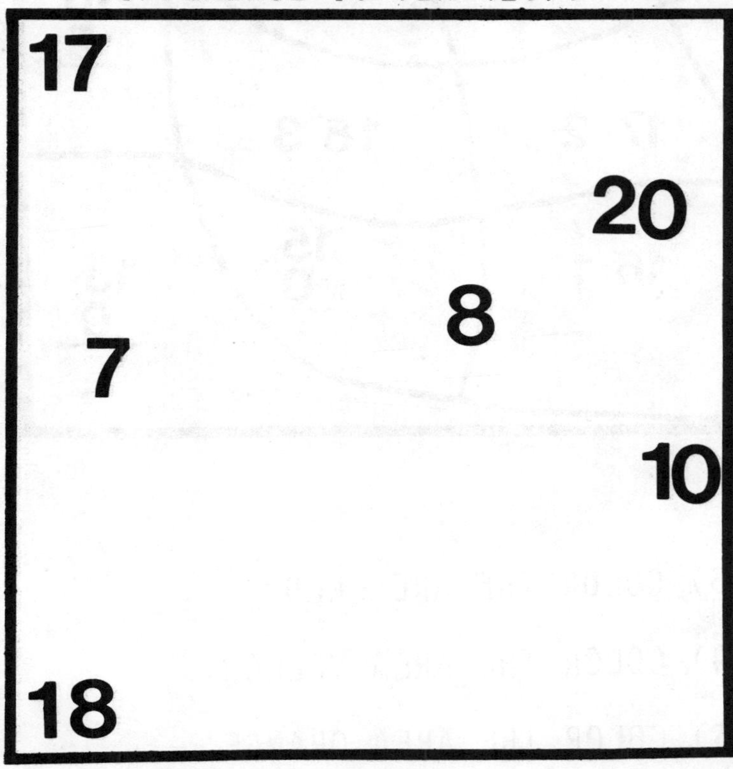

DRAW YOUR OWN LINE PUZZLE. DO NOT PUT IN THE LINES. LET A FRIEND TRY TO SOLVE YOUR PUZZLE.

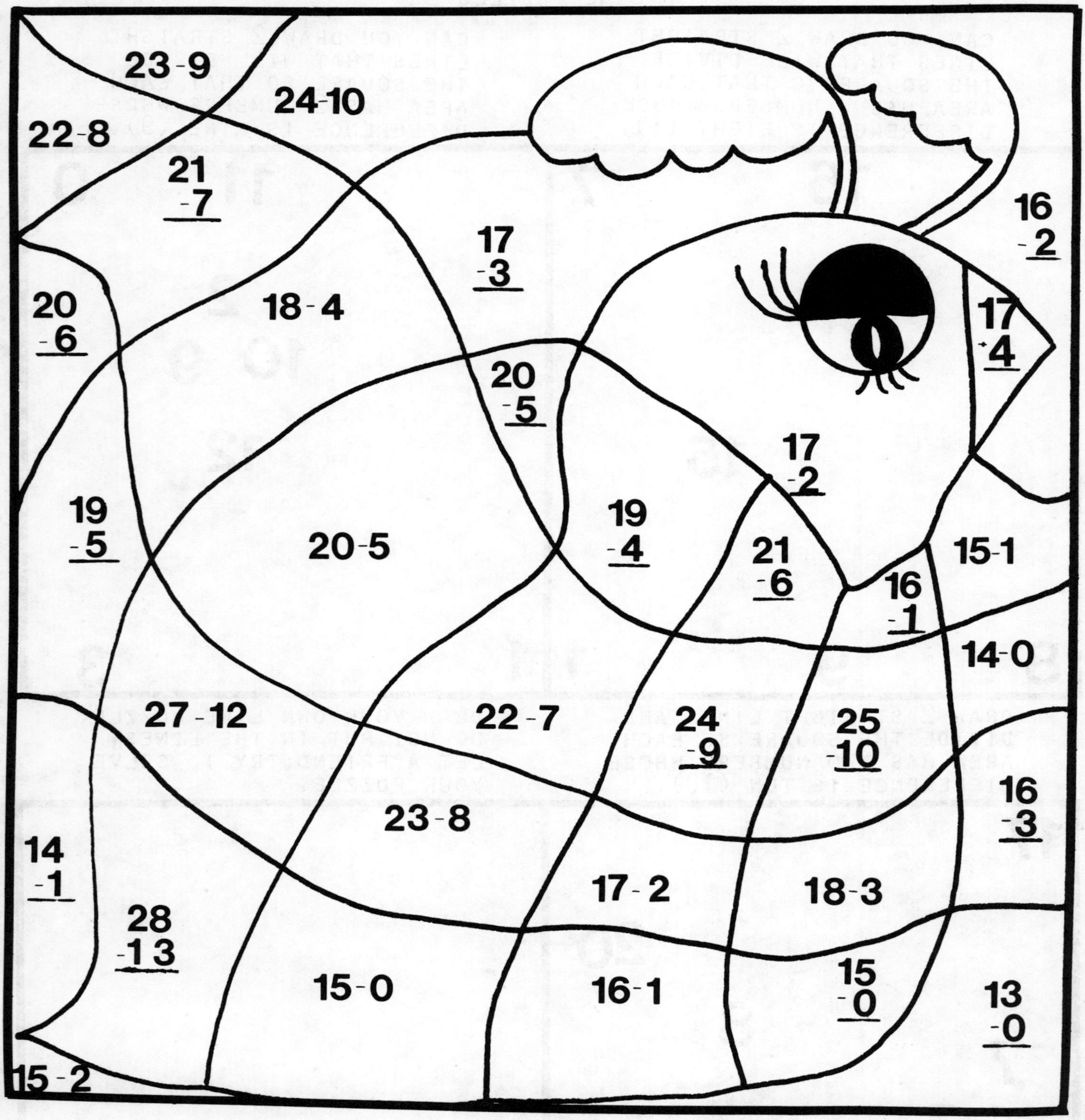

DIRECTIONS:

IF THE DIFFERENCE IS THIRTEEN (13), COLOR THE AREA RED.

IF THE DIFFERENCE IS FOURTEEN (14), COLOR THE AREA YELLOW.

IF THE DIFFERENCE IS FIFTEEN (15), COLOR THE AREA ORANGE.

CIRCLE THE SUBTRACTION PROBLEMS THAT ARE FOLLOWED BY THEIR CORRECT ANSWER.

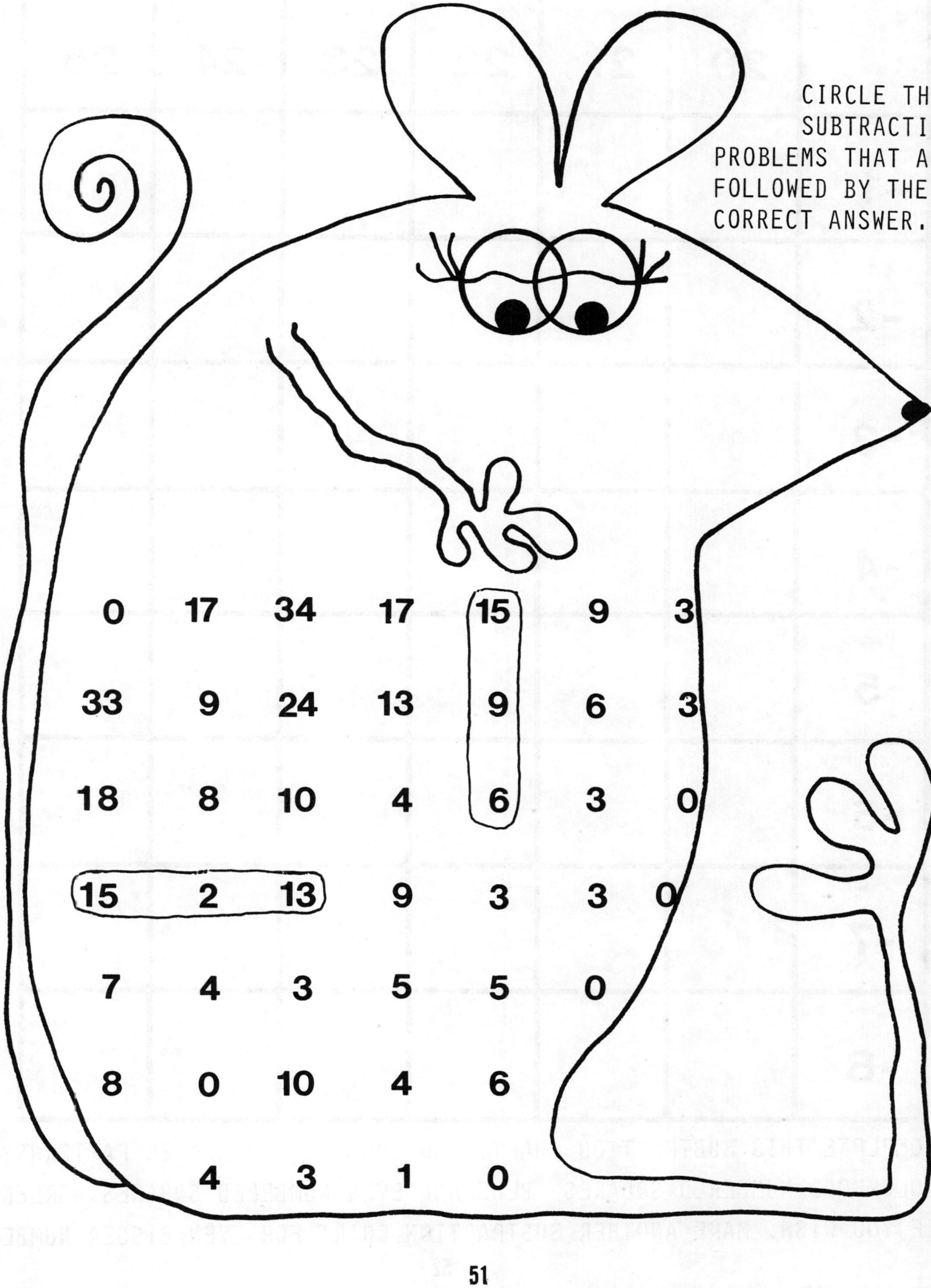

0	17	34	17	15	9	3
33	9	24	13	9	6	3
18	8	10	4	6	3	0
15	2	13	9	3	3	0
7	4	3	5	5	0	
8	0	10	4	6		
	4	3	1	0		

	20	21	22	23	24	25
-1						
-2						
-3						
-4						
-5						
-6						
-7						
-8						

COMPLETE THIS SUBTRACTION CHART. DO YOU SEE ANY NUMBER PATTERNS? COLOR ODD NUMBERED SQUARES BLUE, AND EVEN NUMBERED SQUARES, GREEN. IF YOU WISH, MAKE ANOTHER SUBTRACTION CHART FOR EVEN BIGGER NUMBERS.

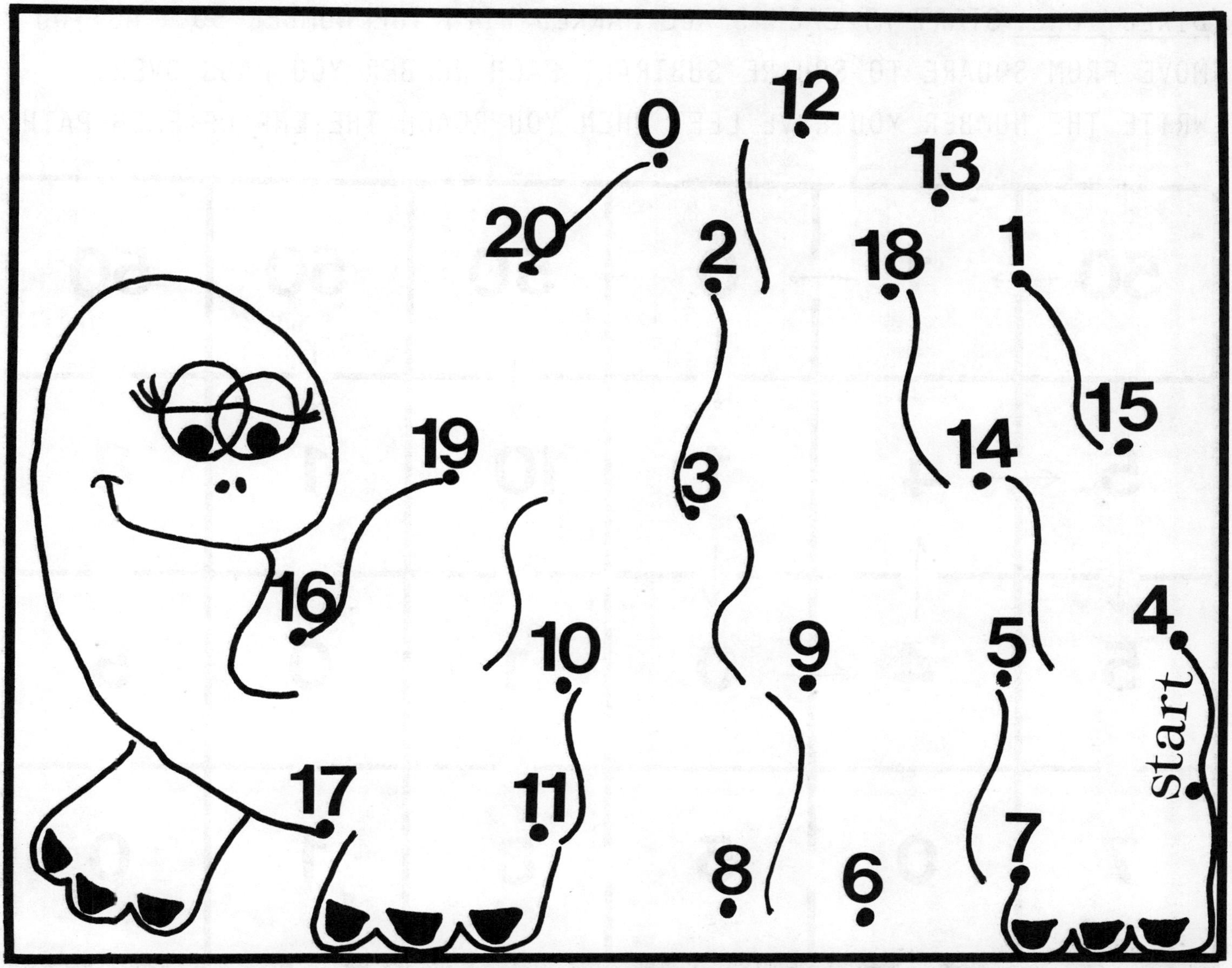

DIRECTIONS: SOLVE THE PROBLEMS BELOW AND THEN CONNECT THE ANSWERS TO COMPLETE THE PICTURE ON THE TOP PART OF THIS PAGE.

1. 10 - 3	8. 9 - 0	15. 29 - 9
2. 9 - 3	9. 8 - 3	16. 5 - 3
3. 9 - 1	10. 9 - 5	17. 20 - 2
4. 13 - 2	11. 20 - 5	18. 4 - 3
5. 20 - 3	12. 20 - 6	19. 20 - 7
6. 20 - 4	13. 7 - 4	20. 15 - 3
7. 10 - 0	14. 20 - 1	21. 0 - 0

DIRECTIONS: START AT EACH SPACE MARKED WITH THE NUMBER 50. AS YOU MOVE FROM SQUARE TO SQUARE SUBTRACT EACH NUMBER YOU PASS OVER. WRITE THE NUMBER YOU HAVE LEFT WHEN YOU REACH THE END OF EACH PATH.

50	1	0	50	50	50
5	4	1	10	1	8
5	4	0	1	0	5
7	0	5	8	1	0
1	4	1	8	0	1
11	0	1	0	1	5
			5	5	

SUBTRACTION CROSSWORD

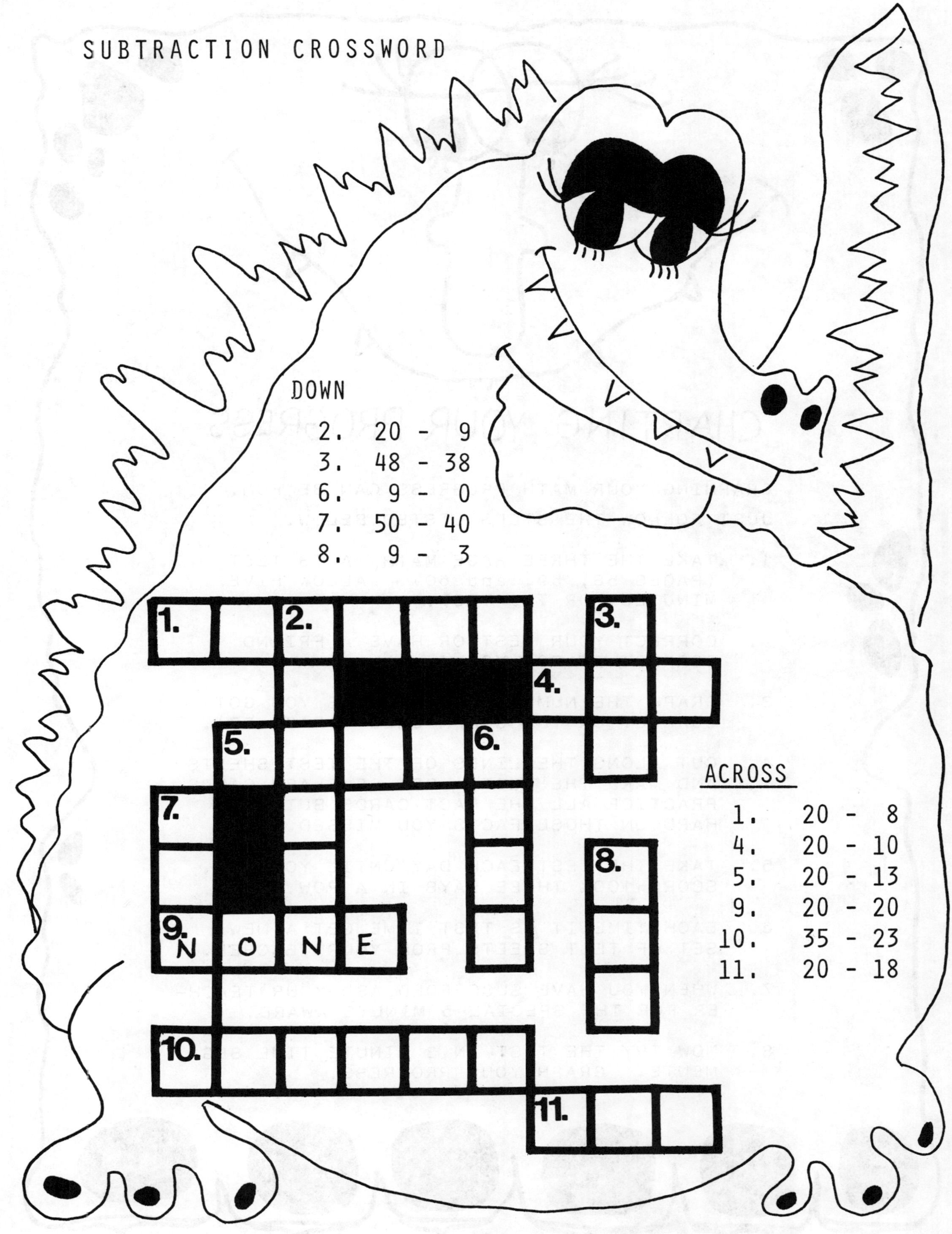

DOWN

2. 20 - 9
3. 48 - 38
6. 9 - 0
7. 50 - 40
8. 9 - 3

ACROSS

1. 20 - 8
4. 20 - 10
5. 20 - 13
9. 20 - 20
10. 35 - 23
11. 20 - 18

CHARTING YOUR PROGRESS

GRAPHING YOUR MATH PROGRESS CAN BE FUN. JUST FOLLOW THE STEPS LISTED BELOW.

1. TAKE THE THREE PAGE MATH FACTS TEST (PAGES 58, 59, and 60). ALLOW FIVE MINUTES FOR THE TEST.

2. CORRECT YOUR TEST OR HAVE A FRIEND CORRECT IT FOR YOU.

3. GRAPH THE NUMBER OF PROBLEMS YOU GOT CORRECT ON THE GRAPH SHEET (PAGE 57).

4. CUT ALONG THE LINES OF THE TEST SHEETS AND MAKE THE TEST A SET OF FLASH CARDS. PRACTICE ALL THE FACT CARDS BUT WORK HARD ON THOSE FACTS YOU MISSED.

5. TAKE THE TEST EACH DAY UNTIL YOU CAN SCORE 100% THREE DAYS IN A ROW.

6. EACH TIME IT IS TEST TIME GET A NEW SET OF TEST SHEETS FROM YOUR TEACHER.

7. WHEN YOU HAVE SUCCEEDED, ASK YOUR TEACHER FOR THE SPECIAL 5 MINUTE AWARD.

8. NOW TRY THE TEST IN 3 MINUTE TIME SEGMENTS. GRAPH YOUR PROGRESS.

NUMBER CORRECT	10	15	20	25	30	35	40	45	50	55	60	63
DATE												
DATE												
DATE												
DATE												
DATE												
DATE												
DATE												
DATE												
DATE												
DATE												
DATE												
DATE												
DATE												
DATE												

SUBTRACTION TEST

THIS IS THE FIRST PAGE OF A THREE PAGE SUBTRACTION FACT TEST. THE REST OF THE TEST IS ON PAGES 59 AND 60. WORK QUICKLY, BUT WORK ACCURATELY. WHEN YOU FINISH, YOUR TEACHER HAS THE ANSWERS.

9 – 3	6 – 4	7 – 2
8 – 3	7 – 5	6 – 0
6 – 1	5 – 5	4 – 4
8 – 2	9 – 5	7 – 3
7 – 6	8 – 6	4 – 1
6 – 5	4 – 2	5 – 3
5 – 4	1 – 1	9 – 0

SUBTRACTION TEST CONT.

BE SURE TO TIME YOURSELF ACCURATELY. EACH DAY AFTER YOU COMPLETE AND GRADE YOUR TEST BE SURE TO MARK YOUR SCORE ON YOUR PROGRESS GRAPH.

8 − 8	7 − 4	4 − 0
2 − 0	9 − 8	10 − 1
16 − 5	12 − 11	9 − 1
8 − 4	8 − 5	9 − 6
13 − 6	6 − 3	13 − 4
6 − 2	9 − 2	10 − 9
8 − 1	11 − 0	4 − 3

SUBTRACTION TEST CONT.

WHEN YOU FINISH WITH YOUR TEST AND GRADE IT, YOU CAN TURN THE THREE PAGES OF THE TEST INTO FLASH CARDS. ALL YOU HAVE TO DO IS CUT ON THE LINES AND PRINT THE CORRECT ANSWER ON THE BACK OF EACH CARD.

1 - 0	9 - 7	4 - 0
5 - 1	9 - 9	10 - 7
6 - 6	5 - 0	7 - 0
7 - 1	10 - 5	1 - 0
7 - 7	10 - 6	3 - 1
8 - 7	10 - 8	99 - 0
9 - 4	0 - 0	8 - 0

FANTASTIC!

____

passed the subtraction facts test in less than 5 minutes.

____

signature

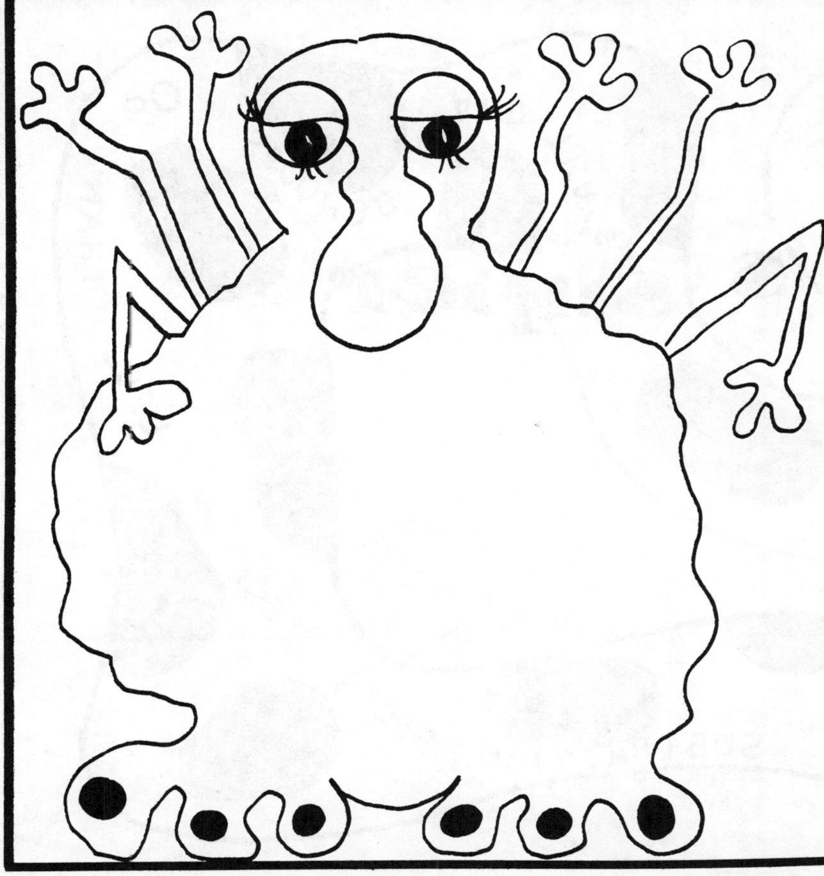

SUPER

____

passed the subtraction facts test in less than 3 minutes

Excellent

☆ YOUR PAPER WAS 100% CORRECT !

☆ YOUR PAPER WAS VERY NEAT

☆ YOUR WORK IS MUCH BETTER

☆ YOU ARE LEARNING THE BASIC FACTS.

☆ ______________________________

____
Signature

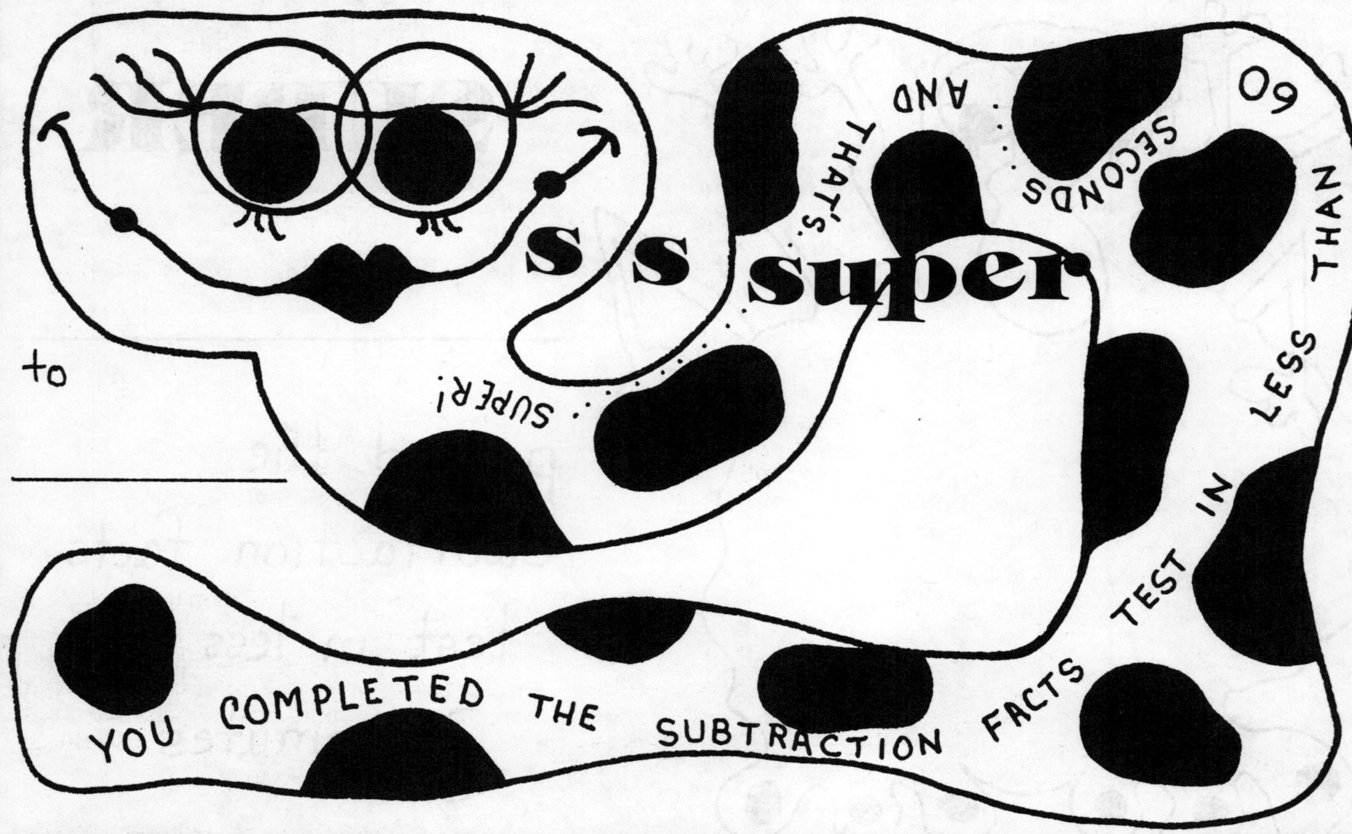